MOBILITY

As everything from immigration, airport security and road tolling becomes headline news, the need to understand mobility has never been more pertinent. Yet 'mobility' remains remarkably elusive in summary and definition. This introductory text makes 'mobility' tangible by explaining the key theories and writings that surround it. This book traces out the concept as a key idea within the discipline of geography as well as from the wider subject areas of the arts and social sciences.

The text takes an interdisciplinary approach to draw upon key writers and thinkers that have contributed to the topic. In analyzing these, it develops an understanding of mobility as a relationship through which the world is lived and understood. The book is organised around themed chapters discussing – meanings, politics, practices and mediations. The book identifies the evolution of mobility and its implications for theoretical debate: these include the way we think about travel, embodiment, and issues such as power, feminism and post-colonialism. Important contemporary case-studies are showcased in boxes. Examples range from the mobility politics evident in the evacuation of the flooding of New Orleans, xenophobia in Southern Africa, motoring in India, to the new social relationships emerging from the mobile phone. The methodological quandaries mobility demands are addressed through highlighted boxes discussing both qualitative and quantitative research methods.

Arguing for a more relational notion of the term, the book understands mobility as a keystone to the examination of issues from migration, war and transportation; from communications and politics to disability rights and security. Key concept and case-study boxes, further readings, and central issue discussion allows students to grasp the central importance of 'mobility' to social, cultural, political, economic and everyday terrains. *Mobility* also assists scholars of Geography, Sociology, Cultural Studies, Planning, and Political Science in understanding and engaging with this evasive concept.

Peter Adey is Lecturer in Cultural Geography at Keele University. His research interests concern the intersecting cultures of mobility, security, design and air travel. His next book *Aerial Geographies* is forthcoming in 2010.

Key Ideas in Geography

SERIES EDITORS: SARAH HOLLOWAY, LOUGHBOROUGH UNIVERSITY AND GILL VALENTINE, SHEFFIELD UNIVERSITY

The *Key Ideas in Geography* series will provide strong, original and accessible texts on important spatial concepts for academics and students working in the fields of geography, sociology and anthropology, as well as the interdisciplinary fields of urban and rural studies, development and cultural studies. Each text will locate a key idea within its traditions of thought, provide grounds for understanding its various usages and meanings and offer critical discussion of the contribution of relevant authors and thinkers.

Published

Nature
NOEL CASTREE

City
PHIL HUBBARD

Home
ALISON BLUNT AND ROBYN DOWLING

Landscape
JOHN WYLIE

Mobility
PETER ADEY

Forthcoming

Migration
MICHAEL SAMERS

Scale
ANDREW HEROD

Rural
MICHAEL WOODS

Mobility

Peter Adey

LONDON AND NEW YORK

First published 2010
by Routledge
2 Park Square, Milton Park, Abingdon, Oxon OX14 4RN

Simultaneously published in the USA and Canada
by Routledge
270 Madison Ave, New York, NY 10016

Routledge is an imprint of the Taylor & Francis Group, an informa business

Typeset in Joanne and Scala Sans by
Book Now Ltd, London
Printed and bound in Great Britain by
CPI Antony Rowe, Chippenham, Wiltshire

British Library Cataloguing in Publication Data
A catalogue record for this book is available from the British Library

Library of Congress Cataloging in Publication Data
Adey, Peter.
Mobility / Peter Adey.
p. cm.—(Key ideas in geography)
Includes bibliographical references and index.
1. Human geography. 2. Migration, Internal—Social aspects. 3. Social mobility. I. Title.
GF41.A34 2009
304.8—dc22 2008054931

ISBN10: 0–415–43399–1 (hbk)
ISBN10: 0–415–43400–9 (pbk)
ISBN10: 0–203–87548–6 (ebk)

ISBN13: 978–0–415–43399–0 (hbk)
ISBN13: 978–0–415–43400–3 (pbk)
ISBN13: 978–0–203–87548–3 (ebk)

For us

Contents

List of figures ix
Key ideas boxes xi
Case study boxes xiii
Mobile method boxes xv
Preface xvii
Acknowledgements xix

1 Introduction 1
Introduction 1
Mobility everywhere 5
The approach 12
Mobile relationality 17
Conclusion 31

2 Meanings 33
Introduction 33
Meaningful mobilities 34
Figures and metaphors of mobility 39
Conclusion 81

3 **Politics** 83
Introduction 83
The politics of mobility 85
Entanglements of mobility 105
Mobile politics 117
Conclusion 131

4 **Practices** 133
Introduction 133
Doing mobility 134
Practice, performance and more-than-representational mobilities 137
Seeing, sensing, moving: examples of mobile practice 150
Motion and emotion: affect and the feeling of mobility 162
Conclusion 173

5 **Mediations** 175
Introduction 175
Mediated mobile societies: auto- and aeromobilities 177
Diffusion 187
Mediating between 196
Augmentation 209
Conclusion 223

6 **Conclusion** 225

Bibliography 228
Index 254

Figures

2.1 Abstracted mobility 36
2.2 Mobility in context 37
2.3 Time–space path visualization 52
2.4 A schematic of Tuan's theory of place 54
2.5 Filipino translocal subjects 79
3.1 The 'tunnel effects' of 'hub and spoke' infrastructural networks and elite helicopter travel in São Paulo 98
3.2 Space–time prisms 100
3.3 Motility and mobility 102
3.4 Inclusive mobility design 112
3.5 The forced displacement of Rwandan refugees 121
3.6 Parkour bodies 126
5.1 Doing office work on the motorway 181
5.2 European business connectivities 181
5.3 Oystercard 'pay-as-you-go' travel in London 216

KEY IDEAS BOXES

1.1 Mobility and moorings 21
2.1 The production of mobilities 35
2.2 Movement, paths and space–time routines 50
2.3 Nomad science 59
2.4 Botanizing on the asphalt, Walter Benjamin, *flânerie* and the arcades 64
3.1 The politics of mobility and power geometries 91
3.2 Motility and mobilities as capital 99
3.3 Flexible citizenship 106
4.1 Phenomenology and the mobile 'body subject' 138
4.2 Mobility and the *habitus* 140
4.3 The tourist gaze 153
4.4 The arabesque of sociality 169
5.1 Transport geographies 178
5.2 Landscapes of global cultural economies 188
5.3 Angels and parasites 197
5.4 The politics of comfort 205
5.5 Technological mobilities, networked infrastructures and information communication technology 210
5.6 The network society 212
5.7 Mobilities and surveillance studies 217

Case study boxes

2.1	Spatial science and the Cold War	48
2.2	Citizenship and sedentarism in Southern Africa	55
2.3	Strategic points and flexible lines – the geometry of the Israeli–Palestinian conflict	61
2.4	Intensities of staying at Henry Parkes Motel	77
2.5	Putting place on the line, Filipino translocalities in Hong Kong	80
3.1	The evacuation of New Orleans	86
3.2	The stratification of helicopter travel in São Paulo	96
3.3	Waiting for the bus	111
3.4	Dams and displacements	116
3.5	Bulldozers, bombs and bridges	121
3.6	Packs, swarms and styles of protest in Khatmandhu	123
3.7	Parkour, utopia and the performance of resistance	126
3.8	Mobilizing the strike	130
4.1	Representing and regulating mobility in modernity	147
4.2	Ascending Glastonbury Tor	158
4.3	Moving together in time	167
4.4	The affective movements of clubbing	171
5.1	Spaces of Indian motoring	185
5.2	Skateboarding and mediation	201
5.3	Waiting for the lift	208

Mobile method boxes

1.1	Rhythm and cultural analysis	30
2.1	Excavating and reconstructing mobilities	43
2.2	Follow the thing	53
2.3	*Flânerie* as mobile ethnography	66
2.4	Mobile ethnography and multi-sited research	70
2.5	Emplacing migrant subjects	78
3.1	The primacy of position	93
3.2	Multi-methods and the slow food movement	128
4.1	The representation of research	145
4.2	Time–space collage	156
4.3	Talking while walking-with	161
5.1	Televisual mobilities and the mobile gaze	193
5.2	Tracing mobilities	219

PREFACE

> [T]here is nothing more that I can do to help it. The board is set, and the pieces are moving.
>
> (Tolkein, *The Lord of the Rings*, 1956)

Is this a book about mobility or a book about ideas about mobility? Essentially it is both. This book will explore a series of different *understandings* of mobility as well as different *approaches* to it, whilst it will set about exploring how exactly mobility *works* and what it means for contemporary issues as diverse as globalization to disability politics.

Before I go into this any further, I will pin myself down to one inescapable truth of mobility: in all its various guises, definitions, approaches, from the most abstract understanding, mobility, at least for me, is a relation. In fact, borrowing from Lois McNay (2005: 3–4) I reckon it is a *lived relation*; it is an orientation to oneself, to others and to the world. Just as Nigel Thrift (1996) outlines mobility as a particular 'structure of feeling', the mobility of something moving through space seems to provide a very certain kind of position, standpoint or way of relating – it is a way of addressing people, objects, things and places. It is a way of communicating meaning and significance, while it is also a way to resist authoritarian regimes. It is the predominant means by which one engages with the modern world. It is a way to bond with one's friend, while it could be the means to threaten a boundary. In a certain sense then, mobility appears much like a notion such as space or time. It is ubiquitous; it is everywhere. It might even be found in everything. But importantly it is almost always born in relation to something or someone.

Mobility is a way of having a relation to, engaging with and understanding the world *analytically*. Mobility does not just exist *out there* for academics, students and researchers to examine from a distance, but it also exists in the pages of books, journal articles and reports; it is rendered in thought and in the imagination. Mobility is a notion. Mobility is a concept; it is conceived. Clearly these imaginations matter for many of the relations mobility is involved in, but mobility also does work *for* us as academics, students and scholars of such a subject. It helps us to investigate processes like globalization, migration, tourism, homelessness, security and transport, from the scale of the international flows of aircraft that criss-cross the globe, to the micro-bodily motions of someone dancing or carrying a bucket to a well. Such approaches are underpinned by particular values and opinions about the world and how it works.

ACKNOWLEDGEMENTS

This book has been born out of a thousand conversations with friends and colleagues who have inspired me along the way. I owe so much to the support of Aberystwyth geography department where the seed of an idea for the book came from. Tim Cresswell's *Geographies of the American West* module I undertook as a second year undergraduate first got me into mobility and Tim provided continued encouragement and advice through my PhD and after. Once office mates Paul Bevan and Suzie Watkin, Aber coffee room colleagues Pete Merriman, Jon Anderson, Chris Yeomans, Gareth Hoskins, Jo Maddern, Deborah Dixon, Rhys Jones, Mark Whitehead and my post-doc mentor Martin Jones, also shaped many of the thoughts that ran through my post-grad and post-doctoral work into this text. At Keele discussions with Peter Knight, Beth Greenhough, Deirdre McKay, Luis Lobo-Guerrero, Steve Quilley and Andy Zieleniec helped hone the initial proposal and later ideas. Tony Phillips accommodated me the time to write, while Rich Waller, Zoe Robinson and Rob Wassell offered the space of accommodation when I needed it! Thanks to James Kneale, the geography department at UCL put me up for a semester where the final touches were made. And the manuscript benefited a lot from Alan Latham's comments. Other mobile colleagues I have encountered along the way include Ben Anderson, David Bissell, Lucy Budd, Steve Graham, John Horton, Phil Hubbard, Pete Kraftl, Mark Salter, Mimi Sheller and John Urry. Students passing through my Mobile Geographies module helped immensely to provide useful thought experiments and case study nuggets.

I want to thank Gill Valentine and Sarah Holloway for taking a chance when they asked me to write the initial book proposal in the first place. Andrew Mould and Michael P. Jones at Routledge added substantial editorial support and patience. In that process, four readers helped me to hone and shape the text into something a lot better. Matthew Tiessen generously allowed me to use his rich and vibrant image for the front cover of this book, providing bucket loads of inspiration when I needed it.

Several authors allowed me to reproduce their photos and images which I am really grateful for, these included Saulo Cwerner, Deirdre McKay and Steve Saville. I am very grateful to SAGE, CORBIS; and Transport for London for Figures 3.5, 5.1 and 5.3. Figure 3.4 is reproduced from the Department for Transport's publication *Inclusive Mobility* (2005).

Thanks to my Mum and Dad for never encouraging me to get a proper job! And lastly to Hayley for always being-mobile with me.

1

INTRODUCTION

[T]he world is a flux of vectors.

(Alliez, 2004: 2)

INTRODUCTION

We simply cannot ignore that the world is moving and, maybe, the world is moving a bit more than it did before. We might even say that mobility is ubiquitous; it is something we do and experience almost all of the time. For Nigel Thrift (2006) even space itself is characterized by this mobility and movement, 'every space is in constant motion,' he writes. Perhaps mobility is not something 'very new' as Anthony Giddens (Giddens and Hutton 2000: 1) comments on globalization, but certainly something 'new' is happening in the world. Aiwah Ong (2006) writes how mobilities have 'become a new code word for grasping the global' and the new and extensive ways in which we live. Without mobility we could not live. Without mobility we could not get to work or to the nearest source of food, neither could we stay healthy and fit. We could not make and sustain social relationships and we could not travel to far off or nearby destinations.

There are a host of statistics and impressive sums to convince you of this. For instance, we know that there are around 200 million international migrants living in the world today marking what former UN Secretary General Kofi Annan described as a 'new era of mobility'. Tourism is widely pronounced as the biggest industry in the world, generating well over $750 billion worth of international tourism receipts in 2005, which has reached around $900 billion at the author's time of writing (earthtrends.wri.org). The industry is worth some $8 trillion or 5.2 trillion euros and employs around 240 million people a year (World Travel and Tourism Council). In Europe this saw over 450 million arrivals in 2007 (European Travel Commission). Migration and tourism are of course supported by transportation – a medium of mobility. In the United States some 675 million passengers flew during 2008. And the travel to work – the commute – has become more and more geographically dispersed. In the United States again, this has seen commuters spending more than 100 hours per year travelling to their place of employment (United States Census Bureau, 2005).

Many of us don't need to see these to be persuaded like this because you can probably think about the ubiquity of mobility in the context of your own everyday lives. Consider getting up in the morning (and I appreciate this is drawn from the account of a male, White, middle-class academic). One wakes up. One gets out of bed, involving the movement of various limbs, a physical displacement from the bed to the bathroom. Skipping a few steps, as part of my work as a university lecturer I must go to work, which involves several other kinds of mobility. I must walk out of my building, using the stairs on the way down, which I might avoid on the way back in the evening by taking the elevator, a form of vertical mobility. I reach the train station after crossing several busy roads. I've had to walk up past the Walker Art Gallery, Central Library and Museum in Liverpool's city centre; their grand architecture reminds me how the place was built on the back of mobilities. Liverpool's earlier wealth was accumulated by the movements of slaves, goods and migrants through the city's famous port. After crossing the busy traffic, I reach Lime Street Station and catch another form of mobility: the train. My mediated journey out of Liverpool is an immediately interesting one as I cross the Runcorn Bridge out of Liverpool; the train route runs parallel with the road along which busy commuters and others move. In the sky overhead an aircraft moves on its way to or maybe from Liverpool John Lennon Airport. In my relatively immobile state on the train, I am able to work, receive text messages, edit this manuscript. At other times I just sit back and try to relax. I get out at Crewe Station and take yet another form of transportation by catching the bus down to Keele University a few miles away.

Let's stop now because I could go on and on, and my recent migration to another part of the country has made this journey now out of date, but this is my point. In just a few hours of a normal day, I have been continuously mobile. I have travelled about 60 miles. My body has performed various mobile tasks. Various mobilities have passed in and out of my body; I missed out mentioning the intake and exhalation of air and food. And in the space of two years I have moved five times. Put simply, in order for me to live, to work, I must move. My body must be mobile and things must be mobile around me, for me. From the water that services my apartment, to the public transport I connect with and use, to the signals my mobile (cell)phone sends out and receives, to the students I will meet later in the day, I expect certain mobilities to synchronize with me.

Just starting with me – my life – my world of habits and routines appears to be fundamentally dependent upon mobilities. In order for me to conduct my career as an academic geographer, to sustain my relationship with my soon-to-be wife, to keep up with my friends dotted around the country and others, I must be mobile in order to achieve the sorts of socialities that compose my day-to-day living and more. Of course, someone else's life will be made up of very different kinds of mobility, some much more extensive and others much more bounded. Some require much more hardship and exertion, and others are far smoother and easier. In this sense, although we may be always on the move, we are also always *differently* mobile.

And while we move ourselves things must be mobile for us too. Many people need to be supplied and provided with services, information, capital or goods. How reliant are we on mobilities? How do we depend on the infrastructural material mobilities of gas and electricity and especially water for energy, agricultural and personal domestic needs. About 23 million Californians depend upon one of the biggest systems of water conveyance in the world, encompassing an enormous infrastructure of pumps, canals, tunnels and pipelines that capture, store and move water, delivered by the State Water Project (Worster 1992). In China, hydroelectric dams form around 16 per cent of the country's total electricity generation. Indeed, dam construction often *moves* people on through forced displacement (see Chapter 4). Of course, the difficulties in supplying and moving clean water is one of the biggest causes of the most desperate cases of disease, starvation and death in the world today.

From fluids and flows to the immovable matter of stone – take how many of the enormous forms and symbols we recognize in the world – all are built upon mobilities. The skyscraper could only be made possible – it could only be enabled – by the ability for people to move efficiently in between floors. The first reliable braking system invented by Elisha Otis in 1853 made skyscrapers a reality once people could move between floors without strenuous exertion or fear (Goetz 2003). Without the ability to move vertically, the skyscraper skylines of New York, Chicago, Shanghai and Hong Kong would not be possible. Indeed, the massive fixities in our world could not be made without mobility, embodied, say, within the mobile labour forces that built the most stubborn and enduring objects as the Egyptian Pyramids. Take the Emirate of Dubai. Witnessing some billions of dollars of investment in tourism and business infrastructures in order to shift its economy away from oil production, the vast Burj Dubai skyscraper city, currently the tallest building in the world, is being constructed on the back of an enormous mobile workforce. Some 1.2 million migrant workers have been transported from India, Bangladesh and South Asia to the region to take up temporary employment as construction workers. Undergoing difficult labour conditions for better wages than they would have received at home, these workers subsequently send cash back to the families they have left behind (DeParle 2007).

Mobility, in short, is vital.

It is exactly these sorts of facts and experiences which have led scholars from many different subject areas and disciplines to argue that life occurs, and perhaps increasingly so, *On the Move* (Cresswell, 2006a). So while mobility might be essential to the form of somewhere like a city, or a building, what does it say to the social lives of people inhabiting these mobile worlds. For Sociologist John Urry, it appears as if the social is mobility. In his book *Sociology beyond Societies* (2000) Urry explains how 'material transformations' are 'remaking the "social"'. By those transformations he was thinking of the mobilities of travel, the movements of images and especially information that are 'reconstructing the "social as society" into the "social as mobility"' (Urry 2000: 2; Urry 2007). In other words, mobility composes society. One isn't just mobile with oneself or even other people. As I have said, the world must be mobile too. Our life-worlds are mobile for us, with us, and sometimes they are against us.

This text is inspired and led by the upsurge of interest in mobility which has resulted in what Hannam, Sheller and Urry (2006) refer to as the 'new mobilities paradigm', or a 'mobile turn' (Urry 2007; Canzler *et al.* 2008). Scholars of travel and tourism have recently sought to place 'mobility at the heart of our understanding of tourism' (Michael Hall 2005: 134; Hannam 2008). In criminology, the issue of mobility is similarly taken as a fundamental subject to tackle. '[T]he criminological world is in motion' (2007: 284) writes Katja Franko Aas. Mobility is clearly social and cultural, but it is also political and economic. In a field such as politics, Chris Rumford has explained how a 'new spatiality of politics' consists of 'flows, fluids, networks and a whole plethora of mobilities (Urry 1999)' (Rumford 2006: 160). Indeed, while policy and regulation enable and shape mobilities, policy itself travels and is shared and copied; as Jamie Peck (2003) puts it, policy is 'in motion' (McCann 2008). Today we are perhaps more aware than ever before of the interconnectivities of financial markets. Extensive use of the word *mobility*, like *globalization*, is perhaps the 'evidence of the very changes it describes' (Giddens and Hutton 2000: 1), and the book aims to use the concomitant debates to gain a fuller understanding of mobility as a key component of the world's unfolding. Further, as it does so, we may gain a more comprehensive understanding of the evolution and arrangement of our *ideas* about mobility too.

This book seeks to illuminate how mobility is engaged in many of the major processes at work in the world today. But let us first consider how these mobilities are apprehended and why only *now* are they being recognized in this way?

MOBILITY EVERYWHERE

Henri Bergson (1911/1950) is probably the most well-known philosopher of movement. Exploring the association between mobility, perception and thought he railed against a *snapshot view*. According to Bergson, mobility and any idea of a world-in-process lay below the individual's capability to perceive that reality. Bergson suggested that perception was rather like that of the snapshot photograph, wherein a moving environment is captured and locked down onto an immobile photograph.

For Bergson and others, thinking, writing about and describing mobility was a process of revealing. Bergson was uncovering an underlying reality of the world that was imperceptible to normal experience. The essential ingredient in this work was an increasing scientific and public consciousness of the coeval properties of time and space (Kern 2003). Quests for knowledge of mobility were regularly undertaken in movements of artistic representation and scientific discovery as David Harvey (1989) shows in *The Condition of Post-modernity*. Artistic movements like futurism saw its leader Marinetti strive to portray the small scale – the little bits and pieces of the world in an irritable animatedness. Marinetti sought out 'the infinite smallness that surrounds us'; he took to task 'the imperceptible, the invisible, the agitation of atoms, the Brownian movements, all the exciting hypotheses and all the domains explored by the high-powered microscope' (Marinetti cited in Jormakka 2002.). Elsewhere the equation of time, space and representation was elaborated on by Paul Klee. '[S]pace, too', Klee writes, 'is a temporal concept' (Klee cited in Jormakka 2002: 6). Taking space with time meant energizing space in a way not previously witnessed

According to Bergson, the perception of immobility is an illusion, a fiction. In reality a body or an object is always moving, it is always changing. He rejected the idea that there can ever be immobility or anything such as a form and even an object, 'there is no inert or invariable object which moves: movement does not imply a mobile'. As he puts it, 'form is immobile and the reality is movement' (Bergson 1911: 302). Similar arguments have been made long after Bergson, particularly by geographer Doreen Massey. Like Bergson, Massey's (2005) argument picks up on the limitations of how we represent a transitory world. Representations such as texts, whose properties 'necessarily fix', appear to deaden a life in flow (to paraphrase Massey). This kind of problematic has been repeated frequently, even by early geographers such as Walter Christaller who bemoaned the 'consideration of the moment, a snapshot of the existing world in continued change'. Like Bergson and others, immobility is presented again as some sort of lie, 'the stationary state is only fiction, whereas motion is reality' (Christaller 1966: 84).

Little mobilities

Amazingly, those curious enough to tackle these questions have come to similar conclusions: immobility or non-movement is explained away as a practical impossibility. Fixity is often seen as a kind of illusion whereas mobility is the truth. Furthermore, finding mobilities seemed to require digging below the scales at which human perception occurred – signified by Marinetti's mention of the microscope. Often imperceptible mobilities have been taken at the scale of the small and micro-scopic.

Greek Epicurean thinker Lucretius, perhaps one of the unacknowledged fathers of the mobilities turn, described the material mobile composition of just about anything. In his *On the Nature of the Universe*, Lucretius (1951) saw or rather imagined the world in a mechanistic and atomized way. The world was in a state of constant mobilization – a vortex of mobile bodies. If something appeared to be immobile it was almost certainly an illusion. For even while an object looked static, it was made up of hundreds or thousands of little pieces which were continuously oscillating. Lucretius renders a world similar to one of those plastic-ball filled play-rooms; hundreds, thousands, millions, billions of ball-like things swirling around, which are moved and displaced as *we* move around too. And yet, we are almost never aware of these mobilities.

For Lucretius, their invisibility depended upon how one was placed in relation to them. Taking the example of sheep viewed from a distant hillside he concluded that the animals' movements were often reduced to an immobile and stationary patch. In observing the example of soldiers in warfare, of wheeling horsemen who 'gallop hot-foot across the mist of the plain, till it quakes under the fury of their charge', their intense movements, viewed from the 'vantage-ground high among the hills . . . appear immobile – a blaze of light stationary upon the plain' (1951: 70). The human bodies he described were considered just as mobile as the world around them. The legionaries and horsemen were mobile not only in the sense of how their own bodies moved through the world as a complete whole, but how their bodies were composed of many moving parts. Jump forward many hundreds of years and we may find very similar arguments made by a series of influential thinkers from ecology and philosophy.

George Zipf's writings spanned across the physical, natural and social sciences drawing together notions of energy conservation, ecology and human behaviour that were extremely influential for geographers. Imagining people as paths (a trope that figures very strongly in our analysis of mobile metaphors in the following chapter), Zipf figured people as temporary agglomerations of matter and energy moving through the body. Taking the imaginary person *John*, Zipf argues how 'John is a *set of paths*' he is also 'a unit' that '*takes paths*' (Zipf 1949: 10–11). There is therefore no John, if John were a self-contained unit, 'there is nothing in this transient matter energy that can be called permanently "John"' (1949: 12). This is because the things that make up John are always mobile.

From this point of view John is a series of paths. John is a system through and over which matter-energy moves. The only fixity of this system comes from the stability of the paths that make John up. The paths enact a patterning of mobility – movements of matter which are not random but follow after one another. The paths move too. Zipf writes how 'we know that the particular system of paths of an aged man are far from being the same as his system of paths when he was an embryo' (1949: 12). Furthermore, Zipf's attention to just what John is leads to the realization of John's interconnectivity with his outside. Treating John as an agglomeration of paths engenders the idea of John as merely part of a mobile system through which matter-energy moves. '[N]o self is an island' the philosopher Lyotard would summarise from Zipf's debate (1984). Existing within what he calls a 'fabric of relations' that are now much more mobile than before, 'Young or old, man or woman, rich or poor, a person is always located at "nodal points" of specific communication circuits, however tiny these may be' (Lyotard 1984: 15; see also Urry 2000).

Taken at an even more minute and biological level, Nikolas Rose (1996) writes how such a body-organism's relation to its outside is open to a process of exchange and transfer. 'Organs, muscles, nerves, tracts' Rose describes, 'are themselves swarmings of cells in constant interchange with one another'. The 'linking and detaching, dying, reconfiguring, connecting and combining' (1996: 185) evaporates the status of inside or outside as they become one another. Furthermore, 'brains, hormones, chemical molecules that connect and transform' erode body-boundaries by connecting '"outside" and "inside" – visions, sounds, aromas, touches, collections together with other elements' (Rose 1996: 185).

Insides and outsides, connections and combinations, philosophers Gilles Deleuze and Felix Guattari describe physical entities moving at different speeds coming into contact with one another to form new-born compositions of bodies resolved (Deleuze 1988; Deleuze and Guattari 1988). These kinds of mobilities are perhaps no better symbolized than in something as small as the food we consume.

Big mobilities

The recent awareness of 'food miles' explains the vast distances and complex commodity chains over and through which food travels to our plate. The little things in life are clearly very, very mobile. Yet these little things may move in very big ways. In the United Kingdom, the Department for the Environment, Fisheries and Food announced in 2005 that 25 per cent of all heavy goods movements was due to food (www.defra.gov.uk). In 2002, food transport had produced some 19 million tonnes of carbon just for the United Kingdom. Little movements of food link-up and merge disparate spaces together like ingredients in a recipe. Thus, the food on one's plate is simply a point on a journey of vectors of food-flow from fields, farms and vineyards from a kaleidoscope of places, some very close and others thousands of miles away. The city and the country merge into a metaphorical baked cake, for in other ways these mobilities may add up to bigger things and other spatial formations. Caroline Steel's recent book on cities (Steel, 2008) suggests how the form of the pre-industrial city was shaped by the traffic of food – of cattle lines moving into towns, by rivers of grain and other foodstuffs. Today's movements are more invisible just as they are decentred in out-of-town supermarkets, megastores and distribution centres in a manner which has radically altered the structure of cities and their hinterlands.

In other words, food mobilities express and build up to much bigger movements or spatial fixities and they especially speak to economic transformations and the social reorganization of towns, cities and the countryside. Thus, in considering small mobilities we must also think about the bigger things. What relation does mobility have to the bigger issues of, for instance, society?

Sociologist John Urry has had the strongest voice in this debate, seeking to re-imagine the social from a static and fixed form to societies which are composed of complex mobilities. In Urry's scheme almost every – although specifically Western activity – requires a form of mobility in one way or another. Drawing on a vast terrain of studies of tourism and travel, to transportation and communications scholarship, Urry (2000, 2007) and Vincent Kaufmann (2002) explain that without these kinds of mobility, societies as we know them could simply not function. Business face-to-face meetings would be impossible. Consuming would be nigh on impossible without the means to receive services or products or have them brought to us. Friendship would be hard without physical proximity or mediated contact by communications like the telephone or email. Leisure activities would be difficult without the means to get to a destination, from the Algarve to Thailand. Some leisure activities are a form of mobility, from cycling, to horse riding, to boating, to skiing. These would be impossible without mobility. Obviously, migration cannot occur without mobility (Blunt 2007). One cannot move home or change their location of residence without the ability to move there. Vincent Kaufmann makes the point in his book *Re-thinking Mobility*, in which he proposes to, 'get rid of the very concept of society in order to replace it with an approach based on movement' (Kaufmann 2002: 18).

Thought of in this way, societies are formed by contemporary everyday mobile processes such as globalization – a process of extensive mobilities. As Kevin Robins notes, 'Globalization is about growing mobility across frontiers – mobility of goods and commodities, mobility of information and communications products and services, and mobility of people' (Robins 2000: 195). For Robins 'Mobility has become *ordinary* in the emerging global order' (2000: 196; my emphasis). Even the relative stabilities of communities, associations of friendship, work and leisure that may at first appear to counter any notion of fluidity, as Arjun Appadurai puts it, are 'everywhere shot through with the woof of human motion, as more persons and groups deal with the realities of having to move, or the fantasies of wanting to move' (Appadurai 1990: 297). Indeed, the way we are governed is increasingly splayed out along the lines and rivers of mobility flow. States and supranational institutions work beyond state territories. Mobility, write Jensen and Richardson, is at the heart of new European spatial visions (Jensen and Richardson 2004; Rumford 2008).

These big mobilities allow societies to expand, grow and exist, while they simultaneously bring them closer. Mobilities mean societies are interlinked by the problematic movement of things such as pollutants.

As a global issue, transnational pollution intensifies our awareness of each other and far-off places (Yearley 1995). Furthermore, 'pollution arises from, and contributes to, the world getting smaller, as it were' (2000: 147). Feelings of compression emerge from issues such as air pollution or disasters such as the Chernobyl nuclear disaster of 1986, 'The world is a smaller place than once it was and other people's pollution crowds in on us' (Yearley 2000: 147). Pollution has become a commodity; it is exchangeable for the reason that it can be moved and, thus, Western and developed nations take advantage of poorer nations willing to take on the burden of their waste (2000: 375). What are known as 'waste trade routes' have seen billions of tonnes of sewage, chemicals and fertilizers flow from Western European countries and the United States to Africa and Asia, resulting in the contamination of sites on the Ivory Coast and efforts to stop this trade in the Basel Convention (1989) and its various amendments (see the Basel Action Network: www.ban.org). E-waste – discarded computers and other electronic consumables – are now being traded in the millions of tonnes each year.

From this perspective, 'connectivity' and 'inter-dependency' are keywords that describe global societies tied together in complex and disparate ways. Urban sociologist Manuel Castell's (1996, 1997, 2000) epic investigations of these connections investigated the rise of an albeit uneven 'networked society' in which the 'space of flows' is superseding the 'space of places' (see Chapter 5). Not surprisingly these complex connections have demanded some different ways of understanding such flows that may combine and interact in random patterns and stretch and skew over space in large and extensive ways (Urry 2003).

These are just a few very brief examples of how mobility is a fundamental human (and non-human) process. The processes which make our world work the way it does, however big and small and however imperfectly, from tourism to migration, from transport to communications, at multiple scales and hierarchies, all depend upon mobility.

The rest of this introduction aims to do two main things. First, the following part of the chapter will outline the general approach the book will take before fleshing out its overall structure with detailed guidelines for each chapter as well as an indication of the particular key ideas, case study, and methodology boxes placed throughout the text. Second, the following part of the chapter will then sketch out several important claims for mobility that will serve as key themes meandering their way through the next four chapters of the book. These will address a cross-section of ideas that revolve around how we address mobility as a kind of *relation*. This outlining is important because it not only pre-empts some of the characteristics of mobility you will encounter, but it also helps to envisage the sort of approach the text takes towards different ideas and conceptualizations as well as the methodological approaches taken towards our mobile world.

THE APPROACH

To reiterate, this book is about mobility. It is about mobility as a key component of the world today, while it is also about the various ways scholarship has tried to address and approach mobility as an idea and an empirical object to be studied and investigated. Therefore, while the book attempts to map out a path through all of this work, the way the debate has its own mobility – the way it *moves* on – adds greater depth to our investigation. In the following section we will think about how mobilities almost always involve a kind of transformation of the contexts and spaces they occur through. In just this way, as the book investigates numerous conceptions, ideas and approaches to mobility, we will track the impact of these ideas, conceptions and approaches upon the contextual disciplinary debates and assumptions they move through. Thus, while this text aims to find out more about mobility and how mobility has been treated, it will also shed light upon the particular contexts these debates have occurred within and transformed.

Having said that, the limits of this enquiry could become endless. I have described how mobilities are essential to our understanding of incredibly wide-ranging processes, so we have to ask the question: where do we stop and where does all of this end? Let us consider three points about the scope and limits of the book's approach.

First, in the following chapter we will see that some scholarship treats *movement* quite differently to *mobility*. Similarly, migration is clearly something very different to transportation whereas tourism connotes something very much apart from the waging of war. It is my contention, however, that an idea of mobility underpins many of these concepts and spheres of study. Obviously, there is an argument that treating mobility in this way makes it meaningless if it is applied to everything (Adey 2006b). But I think this misses the point. This book seeks to render how mobility has been understood as an integral and underlying concept within these processes. In this guise, the text is not about endlessly claiming new territories for the study of mobility, but it tries to scope out where and when it has been rendered as important. More fundamentally, it suggests how mobility operates as a key building-block notion. Mobility is an underlying concept as fundamental – but no less contested – as 'space', 'society', 'power', 'city', 'nature' or 'home' (Castree 2005; Blunt and Dowling 2006; Hubbard 2006). In this light, mobility enables the productive juxtaposition and comparison of diverse research themes.

Second, this book is published in the 'Key Ideas in Geography' series. One caveat this presumes is that the book will limit itself to what is thought of as spatial or geographic mobility. This book assumes mobility is about a displacement of something across, over and through space. Geographic mobility may no doubt involve social mobility; this book is all about mobility's social, political, cultural and economic signatures. At its core, however, we will take mobility as still a spatial displacement – whether material, electronic or potential (Canzler *et al.* 2008).

Third, from this point of view, one could be tempted to simply stop at the ends of the geographers' analysis, but where geography ends and another discipline starts is difficult to know. Many would argue that the strength of a discipline like geography is that these boundaries are not policed, and nor should they be. Influence and expertise are sought from elsewhere; discussion and debate percolate forwards and backwards across many disciplinary divides. Consequently, although the text comes from a geographical point of view, it is very much an *outlook*. It tries to take geography as a starting point but not an end point. Geography forms a window from which to look out and see, placing limits on the width and aperture of our gaze, but not its depth.

How the book is laid out

The book has been divided into four main chapters themed on a specific issue of mobility. Each chapter should not be read as particularly distinct, although it is possible to delve in and out of each one.

While a chapter might be primarily themed around the issues its title describes, this does not mean that these themes have nothing to do with one another. On the contrary they are entirely related and interdependent. Splitting the themes up into chapters in this way just allows us to focus on one particular aspect of mobility in turn. Therefore, you would be entirely justified to ask, 'well, surely there are politics of meanings, and how are mediated mobilities practised?' It is not the purpose of each chapter to box off one dimension of mobility from another. In fact it would be easy to argue that mobilities are almost always meaningful, political, practised and mediated. The themes and issues discussed in each chapter will cross over examples and themes in other chapters. Where I haven't signposted these connections, you should be able to spot their relevance.

The book will be organized into the following chapters:

Chapter 2: Meanings

This chapter explores how mobilities and the study and understanding of mobilities are underpinned by specific ideological and discursive meanings, which are not limited to any boundary between both academic and real social worlds. The chapter traces a pathway through the approaches of early mobility studies beginning with geopolitical theorizing, efforts to dispense with meaning during the spatial science of the 1970s, and later approaches more attuned to experience, symbolism and discourse. By examining the underpinning figures, metaphors and meanings of mobility people frequently *live by* (Lakoff and Johnson 1980), the chapter compares and contrasts how studies of mobility have treated and looked at meaning, how their studies are invariably polluted by different meanings and significances, and how these meanings compare with dominant societal ideologies of movement and fixity. Key figures within these debates such as the Parisian *flâneur* and the nomad are discussed.

Chapter 3: Politics

As mentioned above, mobilities are frequently given meaning; indeed it is these meanings that can make a difference to the way they are treated. Chapter three traces out the relationship between politics and mobility. Exploring a variety of work which has associated mobility politics with ideology, power relations, political contestation and violence, the chapter calls upon diverse case-study examples in order to deal with the complex dimensions of the politics of mobility. The chapter begins by setting out several of the dimensions of this *politics of mobility* drawing out facets of ideology, participation and publics, and the differences that construct it, dwelling upon key theorizations of mobility from Doreen Massey to Aiwah Ong. These issues are then examined through a range of examples that include the following: mobility citizenship; efforts to control and regulate mobility; securitization of mobility, to the inequalities of mobility access and inclusion in disability scholarship. The second half of the chapter then dwells upon how mobilities are constitutive of contestation and political violence, teasing out the mobilities of warfare as well as strikes and protest.

Chapter 4: Practices

This chapter attends to the question, just what is left over from a description or representation of mobility? Drawing upon notions of practice, performance and non-representation the chapter explores how mobility is done in ways that may evade the constraints of representational description, analysis and explanation. We first examine various theorizations of the practice and movement of the body as habits and unconsciously performed routines. The chapter discusses the phenomenology of Merleau-Ponty alongside the work of Pierre Bourdieu and the scholarship of performance theorists. Through this discussion, the chapter moves on to explore the more than visually sensual, tactile and multi-sensorial dimensions of mobility as it composes a variety of mobile experiences and processes. Examples range from the physically exertive practices of running and cycling to dance. The final part of the chapter addresses the much-neglected questions surrounding mobility as it relates to feelings and emotions. Attending to emotion and the affective, the section explores the importance of feeling and collective emotions that emerge through mobilities both together and in time. Case studies range from the spaces of clubbing and crowd sociology to military drill.

Chapter 5: Mediations

How are mobilities almost always carried? How do mobilities almost always carry something? Chapter 4 attends to these questions by investigating the role of mobilities in various processes of **mediation**. Mobilities, the chapter explores, often mediate in the sense that they transport other mobilities or are often transported themselves. Indeed, it is argued that mediation is potentially the most powerful property of mobility as people, non-humans and things regularly travel with and transport one another to different places. The chapter focuses in on how the properties of mediation compose and maintain a multitude of different socialities, relationships and events. In times of insecurity, the mediation of mobility such as the diffusion of disease is a key example as is the illicit movement of other objects and things. The chapter examines what efforts are undertaken to secure and make safe infected or other kinds of insecure or risky mobilities by practices of mobile mediation. In other examples, mediated mobilities can mean places pushed together through mediated technological mobilities that compress time and space, from the aeroplane to the telegram. On the other hand, mediation might mean the distancing of relations buffered and insulated by mobilities that 'get between' spaces and people.

Chapter 6: Conclusion

The conclusion works to summarise and tie the dominant ideas, case studies and methods discussed in the book back together.

Each chapter is split into various subject headings, and several boxes feature at the end of each chapter. There are three different kinds of boxes in the book, 'Key ideas' boxes, 'Case study' boxes and 'Mobile method' boxes. This should be fairly self-explanatory, but let me just briefly outline what they are for: 'Key ideas' boxes will focus on a specific idea or concept in special detail. The sorts of ideas to feature in these boxes are those that I, and the field, have determined to be some of the most influential or important in the study of mobility. Boxes will narrow in on these concepts while tying them to the figures, individuals and fields that conceived them. The boxes will also work as an extended glossary that works to explain and draw out the meaning of a key terminology. The sorts of themes to be included in these boxes include the 'tourist gaze', 'nomadism' and 'flexible citizenship' among many others.

The 'Case study' boxes do something a little bit different. These sections demonstrate the kinds of research projects relevant to the study of mobility. Some of these boxes will present quite famous examples that have had a considerable impact upon our ideas of mobility. Others will present diverse, original and more unusual examples that have perhaps lacked significant attention in the literature, or have remained at the periphery. The intention is that these boxes will showcase some of the best research in this area at present, while illustrating to students just how mobility research can take place. It is hoped that ideas for potential dissertation topics and project essays will be sparked by such boxes.

Finally, the 'Mobile method' boxes demonstrate just some of the ways research on mobilities is really being *done*. This is particularly important given the relatively light touch given to methodological issues in mobilities research (Watts and Urry 2008). Illustrating the methodological tools available to researchers in this area, the boxes outline how the attention to mobility may demand new, old, different or modified kinds of methods of data collection and analysis. These may vary in utility or appropriateness according to *when*, for instance, the archaeological study of mobilities is examined, *where*, and, of course, *what* sort of mobilities are being explored. Students should take away the kaleidoscope of methodological approaches open to their study of mobility. Summaries and advice are bulleted in each one of these boxes as 'In practice' guides.

In each of these boxes may be found references for further reading that are either authors cited in the boxes or recommended texts that did not feature but which are well worth reading on the relevant topic.

MOBILE RELATIONALITY

What do we take away about mobility from the discussion so far?

Central to this book is the importance of understanding mobilities relationally. From the short examples developed above, mobilities are positioned in relation to something or somebody, be that governments, organisms, businesses or food. For John Law (1994) mobility, and its opposite, immobility or durability, have little to do with position but are the result of work and effort. He explains, '*Mobility and durability – materiality – are themselves relational effects*. Concrete walls are solid while they are maintained and patrolled' (1994: 102). Following Law, mobilities and immobilities are the 'special effects' of a relation; they are an outcome or an accomplishment. Something like obduracy is then the result of a relation that may *do* a fixity or undo it – mobilize it.

Take another example of this in Sanford Kwinter's (2001) description of a rock-climber finding his or her way up a mountain face. By mobilising the mountain as Kwinter does we see instead that it is the face of the mountain 'whose flow is the most complex'. The mountain face is in a flow of incredible scale consisting of geological processes of upheaval and uplift, deposition and erosion. We should also note that Kwinter is by no means the first to suggest this. Manuel De Landa (1997a) talks of how oceanic crust is under continuous creation and destruction. Yet in relation to the rock climber, the flow of the mountain face is completely unfathomable, undetectable 'from the scale of duration represented by the electrolytic and metabolic processes of muscle and nerves' (Kwinter 2001: 31).

For Kwinter the two flows of the mountain and the climber support one another up in a relational hold. The mountain supports the climber's movements providing fissures 'wide enough to allow the placement of one segment of one finger, and anchored by sufficiently solid earth'. But both mountain and climber are contingent upon each other's movements. The fissure just described is entirely provisional, only allowing the weight of 'eighty pounds of pressure for, say, three seconds but no longer'. Held for a period any longer than that and the unstable flow of the weathered surface, eroded itself by the flows of wind and rain, may give way – giving way to the climbers grip and the rest of their body (Kwinter 2001: 31).

In the above examples, mobility and immobility are understood as an effect or an outcome of a relation – of a position or of effort and pressure. Before we move on to look closer at the different kinds of relation or relationality this book will trace, there are two ways of thinking about this kind of relationality.

1 *Mobility necessarily involves*

To speak of mobility is in fact to speak always of mobilities. One kind of mobility seems to always *involve* another mobility. Mobility is never singular but always plural. It is never one but necessarily many. In other words, mobility is really about being mobile-with.[1] Often a certain kind of relation between mobility and immobility appears to have been posited.

2 *Mobilities are a way of relating*

Second, mobilities are commonly involved with how we address the world. They involve how we form relations with others and indeed how we make sense of this. In this way mobility may mean an engagement with a landscape; it could be deployed as a label to make sense of an act of transgression; mobility may be engaged as a way to govern, or it could be our use of mobility as an analytical concept.

Comprehending mobility in this way has both analytical as well as empirical purchase. In the rest of this discussion, I will sketch out several dimensions of mobility as a relation which I treat as fundamental characteristics for understanding how mobilities are treated, and how they come to constitute the patterning and dynamics of the social world.

Taking place, making space

Imagine you are sitting near a pond or a lake on a hot day. The water is perfectly still and you decide to put your feet in the water to cool them down. What effect will your feet have on the water? Every move you make will conjure ripples that move across the surface of the pond. Even a twitch of your toe can create an effect. Watch your waves disturb the water lilies, how small insects and other pond life might fly away or alter their path across the water. In other words, the mobility of your leg cannot ever be taken apart from the other entities and their mobilities on the pond. Imagine again that you decide to kick your feet out and make a splash. You kick out your right leg and bring it down towards the water. Upon entering the water, the force and mass of your leg displaces the surface of the pond. You temporarily part the water by pushing it out away from your leg that has now taken up its position where water once was. Pond water also enters the air as the pressure of your leg's downwards movement squeezes it outwards and upwards. Almost a fraction of a second later the mass of the pond creates a resultant force that rushes back in to fill the gap your leg has created in the pond's volume area. The meeting of the water over your leg comes at such a speed that it surges together resulting in a satisfying splosh of water – the kind of upsurge that expert divers try to avoid as they enter the pool below.

We might take what has happened here as a useful little analogy of what happens every time we move or are mobile. Space is changed. The subsequent movement of the pond is displaced, charged, sploshed, frothed around due to our mobility. Of course, the space around us and through which we then move is disturbed, but it is also altered for others. For the water beetle, crossing the pond is now a rather different proposition. For someone passing by, observing their reflection in the glassy surface of the pond, we have disrupted their stare. In other words our mobilities make waves.

Massey (2005) articulates a similar idea in an account of her train journey from London to Milton Keynes. This is less about travelling through space or across it, but as she explains it is because 'space is the product of social relations' our mobility helps or works 'to *alter space*, to participate in its continuing production'. In the process of travelling Massey has become part of the making and breaking of links that erodes or reinforces Milton Keynes as an 'independent node of commuting'. The travel to and arriving at Milton Keynes is to do more than travel superficially across or through space, instead 'you are altering it a little' (2005: 118). Discussing another example of her parent's home in the Lake District, Massey imagines the many ways her and her sister's mobilities shape and are shaped by the nearby mountain, which is 'still rising, still being worn down (and the constant tramp of hiking boots, not to mention mountain bikes, is a significant form of erosion in the Lake District), still moving on; my sister and I just here for a long weekend, but being changed by that fact too'.

From this cue we should take notice of Aiwah Ong's (Ong 1999) preference for the term 'transnationalism' to describe the mobilities of international migration and 'flexible citizens'. As Ong explains, '*Trans* denotes both moving through space or across lines, as well as changing the nature of something' (1999: 4). It is these sort of geometrical and transformative relations of the effect of mobility upon contexts, geographies and other people's mobilities, that I consider throughout the text.

Moorings

In his reading of French urban philosopher Henri Lefebvre, John Urry (2003, 2007) posits a mutually beneficial relationship between mobilities and relative immobilities or permanencies that should be vaguely familiar by now. The complex mobilities of social life require being placed in relation to immobilities or what he calls moorings, moorings that are solid, static and immobile. Mobile life has become constituted through '...material worlds that involve new and distinct moorings that enable, produce and presuppose extensive new mobilities' (Urry 2003: 138). Without these immobilities or moorings, there can be 'no linear increase in fluidity' (Urry 2003: 125).

Key idea 1.1: MOBILITY AND MOORINGS

Initially developed in *Global Complexity* (2003) and through his later writings Urry shows how complex mobility systems are supported by immobilities or infrastructural moorings. These may well include vast and heavy facilities and technologies that are far more rooted, but which permit other fluidities. The technology of the mobile phone is an obvious example sustained by various broadcast aerials and relays. GPS navigational systems likewise rely upon an albeit relatively immobile network of satellites in geo-synchronous orbit (see Chapter 5).

However, these fixities are not necessarily permanent or related to an external immobility. Mobile systems can be moored by temporary immobilities of storage such as the 'overnight stay of a car in a garage or an aircraft on an airfield or information within a database or a passenger within a motel' (Urry 2003: 125), often requiring careful coordination and management. On the other hand, storage can give way to even more temporary stages of rest and preparation, such as 'a bus-stop, voice mailbox, passport control, railway station or web site'. (Urry 2003: 125).

Later on in the book we will examine the prescient issue of how certain kinds of mobile people may require the fixities of other people to support them. But this dialectic of fixity and movement is an issue that underpins the dynamics of some of the dominant processes at work in the world today.

Further reading

(Cresswell 2001; Adey 2006b; Urry 2007)

In this light immobilities or fixities act as enablers. They are fixities that seem to permit, provoke or enable other 'mobile machines'; these could be anything from a mobile phone, a car, all of which presume 'overlapping, and varied time–space immobilities' (Urry 2003: 125). If we think of somewhere like an airport, the mobile machine of the commercial jet plane requires the immobility or the mooring of the airport city to provide a temporary physical port for the aircraft to land, refuel, unload and load (Adey 2006b). The immobility of the airport city that employs tens of thousands of workers, further supports this claim (Urry 2003: 125).

This relation is played out time and again. Take the simple registering of mobility itself – the problem of actually discovering mobility in the first place. There is the issue of whether fixity is a necessary evil if we are to actually recognize something as mobile. Hugh Prince (1977) once asked whether we need some kind of fixed reference points, whether we need immobility in order to identify the mobilities that temporality animates:

> We may stand on the bank of a river watching a duck paddling upstream, moving into new water with each stroke, while the current gradually carries it away from us in the opposite direction. In relation to the body of water, the duck sees that it is making headway upstream, but in relation to the river bank it seems to be drifting downstream. To return to its starting point it will have to paddle faster to overcome the speed of the current. In all events, many different kinds of movement are occurring simultaneously. [. . .] In the perspective of time, all places, all positions, all locations move.]
>
> (Prince 1977: 21)

Even while all locations and all positions seem to move, these locations and positions give us very different senses of mobility. As an artist like Paul Klee needed a chair to appreciate a painting, mobilities and immobilities compose different positions from which we make sense of the world and each other.

Such moorings are not necessarily absolutely mobile. Tim Cresswell (2001) discusses the complex and differentiated mobilities and identities of people at Changhi Airport, Singapore, after he took the elevator to the wrong floor. As the elevator's doors opened on the service level of the airport, Cresswell describes what are the 'symbiosis of mobilities'. He writes how the floor was inhabited by 'the people who work there – the people who staff the check-in desks and the people who clean the toilets and empty the bins who come in from the city on a daily commuting cycle' (2001: 23).

The issue is not so much that these employees are immobile; of course, they are nothing but mobile. What is at stake is how these relatively mobile people support and service the mobilities of passengers who pass through the space of Changhi. We, therefore, need to account for the fact the mobilities not only occur in relation to others but that mobilities happen with others in a sense of a symbiotic path dependency – trajectories that intertwine and share a common direction. This is not simply a matter of a single mobility bouncing around and off other movements, resembling the physicalist imaginings embedded within numerous public transportation models and predictions, which we will critique later. What we need to realize is that mobility often occurs with others. Mobility is a social activity of companions of people and things who move with. Some people are dependent on other people in order to move; children might have to travel with their parents or a responsible adult; a mobility-impaired person may be dependent upon a friend or a relative to help him or her get around.

Spatial fixes

So we might need fixities in order to see and recognize mobility. Fixity provides a sort of backdrop for us to distinguish mobilities against. Yet various authors argue that a spatial fix operates within the logic of the most mobile and ephemeral process of all – contemporary capitalism. Geographers such as David Harvey and Neil Brenner have demonstrated that there is more to assumptions of the free-moving fluidity of capital. Harvey posits that the circulation of capital actually requires several phases of fixity construction and deconstruction. The tension is evident in the landscapes of cities which function as 'spatial fixes'. The city is a site in which production and consumption can be located in such a way that capital accumulation will continue. These fixities are only ever temporary states of equilibrium. Crisis points meet the city with creative destruction by restructuring it for the next surge of capital accumulation. In other words, the perpetual running of capital will not occur without fixities around which it can accumulate.

It is recursive too. As mobilities are enabled by fixities, mobilities construct and create further fixities. A spatial fix is required in order to create what Harvey describes as a 'structured coherence to production and consumption within a given space' (1985: 146). These activities are secured by producing territorial or spatial forms and configurations, 'upon, within, and through which expanded capital accumulation can be generated' (Brenner 1998). Later drawing on Whitehead's notions of enduring objects (Whitehead 1979), in *Justice, Nature and the Geography of Difference*, Harvey (1996) shows how these processes always meet crises. At such a point the fixities created by the accumulation of capital melt into thin air, torn down repeatedly like the casinos in Las Vegas. As each round of accumulation reaches crisis point, these spatial fixities are once more swept away to be reconfigured elsewhere, constructing new infrastructures for new accumulation.

As Harvey describes the construction of 'actual permanences' such as cities or other social institutions (Harvey 1996: 81), there are other sorts of fixities that operate at a spatial scale. According to Brenner, scale operates as a territorial entity in order to offer another dimension to that of flat spatial fixities. According to Brenner, capital's need to territorialize and accumulate within fixed and discrete geographical spaces is repeated in a scalar dimension as various geographical scales offer hierarchical and bounded territories for capital accumulation. We can also see these tensions between fixity and fluidity evident within wider systems of exchange that link spaces of accumulation together (Graham 1998: 176).

Like Urry's moorings what we are seeing here is a dialectical feedback loop occurring across both material and social relations as 'spatial arrangements are reciprocally tied to movement processes' (Abler *et al.* 1971: 236). From the physical landscape to complex social and economic relations, mobilities may create structures and fixities which may influence further movement.

Positions

Urry's concern for the relational moorings and stabilities of spatial mobilities resonates not only in the literature of complexity and systems theory, but with the influence of post-colonial, feminist and queer theory. It is at these thresholds that researchers have emphasized the relational and mobile properties of subjectivity and identity. Elspeth Probyn (1996), for example, emphasizes identity as a process of continuous departure. The capacity for one to move in and out of different subject positions depends a great deal on the recognition of multiple identities and an ability to see from one's own perspective as well as others. In this light, the relative fixity of Urry's material moorings are mirrored in this literature's attention to the fixed and fluid positioning of subjects.

Set in the context of the feminist critique of masculine subjectivities and the subversion of Eurocentric narratives of Empire and colonial rule, thinkers such as Donna Haraway and Sandra Harding wrought powerful criticism of a fixed, monocentric, masculine scientific gaze. Positioned in relation to the world as an omniscient all-seeing eye, academic hierarchies of knowledge production were decentred by researchers inspired by this work. In advocating an account of difference, and especially the lives of women, a more mobile gaze was one that could shift, travel and look around towards the *standpoints* and experiences of other marginalized groups. Of particular note are the writings of mobile women travellers, previously distanced from what has counted as scientific knowledge and masculine historiographies of academic study (Blunt 1994; Domosh 1991). American studies author Virginia Scharff writes:

> If we try to see the great events of our history through the eyes of women in motion and action, those events and the places they happened, look different.
>
> [. . .]
>
> These movements have been difficult to see clearly, because historical maps – both graphic and linguistic representations of the ways in which people have marked place – have generally been drawn by, for, and about men.
>
> (Scharff 2003: 3–4)

From post-colonial literary criticism Edward Said's *Orientalism* (1978) as well as *Culture and Imperialism* (1993) suggested the academy's resistance to the 'ravages of imperialism' were born from the same 'exilic energies' as the migrant between 'forms', 'domains', 'homes' and 'languages'. Of course migrancy was an important condition for many post-colonial writers, while it simultaneously reflected their positioning towards the 'unhoused' and 'decentred':

> [I]t is possible, I think to regard the intellectual as first distilling then articulating the predicaments that disfigure modernity – mass deportation, imprisonment, population transfer, collective dispossession, and forced immigrations.
>
> (Said 1993: 332)

This forced one to become more like a traveller, depending on their mobile positioning in order to 'understand a multiplicity of disguises, masks, and rhetorics'. For Said, 'Travellers must suspend the claim of customary routine in order to live in new rhythms and rituals . . . the traveller crosses over, traverses territory, and abandons fixed positions all the time. (cited in Howe 2003).

At the same time, this kind of 'commitment to mobile positioning' is not such a simple task. The fluidity of position making is such that it is often moored by the pull of one's own position. As Haraway explains, 'One cannot "be" either a cell or molecule – or a woman, colonized person, labourer, and so on – if one intends to see and see from these positions critically' (Haraway 1991: 192). In other words, it is not always so easy to step in and out of one's own positioning, a problematic we will discuss later in terms of methodology.

Trajectories and synchronicities

Mobilities inevitably involve moving alongside and synchronizing with one another, and it is because of their relatively similar trajectories and their synchronicity into temporarily stabilized configurations, that they can appear immobile. Henri Lefebvre (2004) articulated this sort of idea as he contemplated the objects sitting in front of him at his desk. At that moment Lefebvre noticed how his table, the pen and other things on the desk seemed so 'inert'. He postured whether their inertness was a product of the fact that the objects had merely similar trajectories to him and, furthermore, that there was little evidence for their activity. The seemingly immobile status of the objects in front of him had hidden a history of trajectories of movement and social relations. As Lefebvre argued, 'this immobile object before me is the product of labour; the whole chain of the commodity conceals itself inside this material and social object' (Lefebvre 2004: 82).

As we unpack the histories of people and objects we can reveal the concealed mobilities hidden within them, often for very particular reasons. But of course, not all things are able to reveal their flows in the same way. Some are much harder to address than others. Urban sociologist Simmel uses the example of the stone and the river to describe nature's much more one-sided expression of its capacity to 'last' and 'flow'. While the stone and the stream are symbols of things with much more obvious trajectories, 'the human being is at the same time always something lasting and something flowing away for our sense' (Simmel cited in Frisby 1985: 115).

There is more to it than this, for the synchronicity of Lefebvre with the items on his desk belies a sense of his closeness to the objects before him. Their ready-to-handness (Heidegger 1977) conceals both the need for proximity and connection in order for social activities and practices to occur and for us to be aware of the mobilities that make them up.

It could be argued that we may be so closely caught up in trajectories of movement and other synchronicities of people and objects that their fluidity is lost. Built architecture provides a useful vantage point as our immediate impression of it exhibits the sort of one-sidedness Simmel finds in the stone. In Stuart Brand's (1994) *How Buildings Learn*, he asks us to 'look up from this book, what you almost certainly see is the inside of a building. Glance out a window and the main thing you notice is the outside of other buildings. They look so static' (1994: 2). In relation to the building we are the moving actants. We move around and abound it while the building remains still. Brand tells us how architecture 'is permanence'. In wider use, the term *architecture* always means 'unchanging deep structure' (1994: 2). For Brand, all this is 'an illusion' reflected if one looks closer at the meaning of the word *building* which Brand takes as both a verb and a noun, 'It means both the action of the verb BUILD and "that which is built."' Therefore, while buildings might appear permanent and immobile, they are 'always building and rebuilding. The idea is crystalline, the fact fluid' (Brand 1994: 2). J. D. Dewsbury similarly highlights how spaces like buildings mean that we are so close to the mobilities of such a structure, that we are unaware of them. Dewsbury writes, how 'whilst you are there it is falling down, it is just happening very slowly (hopefully)'. Therefore, in finding ourselves in a space or a world that is incessantly 'bifurcating and resonating amongst the different movements of its many compositions, our subjectification is always occurring' (Dewsbury 2000: 487).

On the other hand, the desire for closeness and ready-to-handness shows how synchronicity is a necessary component that constitutes social relations and justifies or provokes some of the most extensive mobilities people may make. Harvey Molotch's famous phrase, the 'compulsion to proximity' captures a common societal need for subjects to hook up, connect or present themselves in certain situations and places (Urry, 2002). Urry presents various social, familial, legal, place-based and other sorts of obligations that require the closeness of geographical co-presence. These may require seeing 'the other' 'face-to-face' and thus enabling of 'body-to-body' language and gestures (Urry, 2007). This can be incredibly important in building trust for friendship, working relationships or more intimate loving ones. The synchronicity of 'free time' is regularly important in allowing planned and scheduled encounters away from work or perhaps family life. Obligations to certain places may mean the requisite of co-presence at the same place at the same point in time. This could be as important as an architect's site visit with his or her client, or 'being "by the seaside"' to strolling along a valley. Similarly, event obligations require connection not necessarily with places but with eventive occurrences. This could mean direct co-presence at a live event, such as 'political rallies, concerts, plays, matches, celebrations, film premieres, festivals' (Urry, 2004, p. 32), or it could mean more mediated coordination and connection in order to experience a more distanciated event. Speaking to friends and family during 9/11, or the London bombings of 7/7, and urging them to get to a television are cases in point (Urry, 2004).

Rhythm

Mobilities usually synchronize in rhythmic patterns. If you are a student imagine your daily term-time mobilities or consider your own daily routines or, even more simply, your biological needs. You must move from one place to the next for lots of different reasons, and your mobilities may have no reason. Typically, walking, driving, getting the bus to campus to take your morning lecture would have no purpose if you didn't synchronize your mobilities with other people. Our conceptions of time and timetables allow us to synchronize with one another because we have a relatively standardized notion of what time and day it is. Thus, a student can be confident that their movement to campus will coincide with the movement of their fellow students to meet, sit and work with. Just as a lecturer is confident, or rather hopeful (!), that his or her students will synchronize their mobility into the campus for the beginning of the lecture. Indeed, just getting to the campus will no doubt entail its own kinds of necessary and unwanted synchronizations of movement. One's bus will have to be met at a particular time; therefore, one must ensure their mobility gets them to the bus stop in time for the bus. At the same time, the parallel synchronicity of other mobilities in the morning and after-work creates the congestion of the 'rush hour' traffic jams and public transport overload. Without all these mobilities synchronizing, things would very quickly break down.

It is often when rhythms break down that we become aware of the scale and scope of these mobilities. A breakdown of rhythm highlights mobilities that haven't quite coincided. There are of course far more important political and power relations embodied in this sort of dynamic, which we will think about later, that have something to do with people's restrictions of their own rhythms, to rhythms they cannot ever harmonize with, and to rhythms that may well be invisible.

Our understanding, appreciation and grasp of these rhythms appear to depend an awful lot on us and our own rhythms of moving about. The fact that I am usually in the tutorial room before my students probably means that they are oblivious to my 60-something-mile commute south down the country. Yet the occasionally late student on a Thursday morning gives me an idea of the timing of weekly events on a Wednesday night at the university student union. As Lefebvre (2004: 82) puts it, 'Our rhythms insert us into a vast and infinitely complex world which imposes on us experience and the elements of this experience.' Stepping out of and into different relations of different rhythm enables us to see worlds that are the same yet 'differently grasped' (Lefebvre 2004: 83).

Mobile method 1.1: RHYTHM AND CULTURAL ANALYSIS

Developed initially by Henri Lefebvre (2004), and built-on substantially by authors such as cultural studies writer Ben Highmore, an attendance to rhythm is proposed as a *methodological orientation* to the cultures of city spaces.

Although rhythm analysis is never formalized as a prescribed methodology with specific rules and regulations, for Highmore, foregrounding rhythm has enabled a bridging between scholarship concerned with the physicality of urban spaces, alongside the forceful signifying writings, media and culture that describe and make it. As Highmore puts it, 'If cities have rhythms, so do all accounts of cities: movement is as essential to film, for instance, as it is to the actuality of the street' (2005: 9). Examining rhythm, therefore, highlights the powerful 'descriptive powers of cultural material' (2005: 9) opening up an array of mediatized and meaningful representations of the novel, the film, sound, music and much, much more.

Journeying through the writings of Brooklyn author Paul Auster (1987), for instance, can tell us much about the everyday banality of the flows and fluxes of urban experience. Indeed, the genre of detective feature commonly features mobility as a key element. As Highmore (2005) shows in his book *Cityscapes*, *The Matrix*, along with Marvel cartoon comic-book heroes such as *Spiderman* and D.C. Comic's *Batman*, are exemplary portrayals of the rejection of the material and economic form of the city. Disposed of by the character's hyper-mobility (see Chapter 3) the economic-material organization structure of New York's and Chicago's skyscrapers above and traffic jams below, form not barriers but navigable impediments to be crawled on, leapt over, swung and even flown past (verticality is a strong feature in much science fiction, particularly *Bladerunner* and *Minority Report*). Likewise, social, political and economic logics which have led to particular mobile experiences of verticality (the skyscraper) are reflected in fictive experiences of the horizontal 'centrifugal' expansion of the city into the sprawling metropolis of Los Angeles and the highways that link its parts together (as also shown in the second *Matrix* film and many others).

These evocative portrayals show the body's experience of the ebb and flow of urban existence, telling us something about the past, present and future of mobilities in the city.

In practice

- Textual and discursive analysis of cultural products from film to literature can shed light on the everyday and the extraordinary experiences of mobility, particularly within city spaces
- Examining the rhythm, pace and tone of these texts can reveal the differentiated experiences of urban mobilities
- Science fiction writings and filmography, while reflecting current anxiety and hope, are particularly powerful crystal balls of future urban mobilities

Further reading

(Thacker 2003; Lefebvre 2004; Highmore 2005)

CONCLUSION

This introductory chapter has argued that mobility is a fundamentally important process that underpins many of the material, social, political, economic and cultural processes operating in the world today and past. Life moves in sometimes small and other times very big ways. And thus, if we are to understand this world, from migration to infrastructural services, from disability rights to the driving of cars and the spread of disease, mobility is surely as important to us as the conceptions and debates that surround notions of space, time and power.

The chapter has, therefore, tried to set up some important tools that we can use to unlock and interpret both understandings of mobility and mobility itself as a social problem. Fundamentally, we are taking mobility as a concept that performs and holds together a series of types of relation. We have already examined briefly above how mobility has rhythm and direction; mobilities connect, meet, position and synchronize. In the following chapters we will explore this relationality in more depth through the different modes and properties of mobility, from meanings to politics, to the way it is practised and mediated.

We have, therefore, set up the direction and organization the book will take. The following chapters are thematically organized into ways that tackle the dominant characteristics of mobility and how it might be approached. In the next chapter, we will see how one of the dominant ways mobility has been apprehended is through the sorts of meanings it is given.

NOTE

1 This notion is inspired by a comment from Mimi Sheller at the 2007 Association of American Geographers' Annual Meeting in San Francisco.

2

MEANINGS

> The very vocabulary in which we discuss questions of mobility is [. . .] inevitably value-laden.
>
> (Morley 2000: 41)

INTRODUCTION

Imagine walking through a wood, along a beach, or a street. Consider how these movements could come to mean something. Imagine how they might be interpreted by somebody else walking by. Well, walking down the street during rush hour might connote images of business commuters on their way to work, the corporation, business and capitalism. Walking along a beach might be taken as a sign of leisure, a signal of being on holiday or vacation. It is the kind of image to feature on tourist brochures and guides of destinations such as Spain, the Bahamas, Portugal or Mexico. Alternatively, walking through somewhere like a forest could be interpreted as an escape to nature, a move away from city life or even a way to enjoy a day off work.

What has happened here? The movement of the walker down the street or across a beach is being taken as something more than a physical displacement and an exertion of energy. The walk has had significance ascribed to it. It has been given meaning. For many geographers and social scientists, it is commonly the places in which we live, work or dwell that make a difference – that define certain sets of meanings that are ascribed to mobility. In this light, while mobility is interpreted and read by someone, that person will nearly almost always do so within a wider context of established societal norms, codes of conduct, belief systems and ideologies which may or may not be common to a particular place. Furthermore, while social and spatial contexts make a difference to the meanings we give things, these meanings matter even more vehemently for the way mobilities may be treated.

This chapter seeks to uncover just what sort of meanings mobilities are given and how they are produced through different real-world and academic contexts. The chapter is interested in how particular ideas about mobility are assigned with certain significance and meanings, be they from academics ruminating on transport patterns or the transport planner designing the very same transport system.

The chapter is structured in the following way. The next section of the chapter looks at how mobilities are produced as meaningful through the coinvolvement of social, cultural and political contexts. The following sections then examine how several metaphors and metaphysical frameworks have dominated this production. Explained in more detail later on, key figures or frameworks are used to unpack some of the dominant meanings and ideologies mobilities are given. The metaphysics of nomadism and sedentarism is understood through the geometries of the nodal point and the line connecting it. The final section of the chapter examines approaches which move past the binary divisions of sedentarist and nomadic thought.

MEANINGFUL MOBILITIES

According to geographer and cultural theorist Tim Cresswell (2001, 2006a), mobility without meaning is simply movement. Like the abstract equation familiar to physicists: speed = distance/time, movement is understood as a similar abstraction. Thus, for Cresswell, there is much more to mobility than its usual connotation – movement. Mobility is movement imbued with meaning. The way movement gains meaning and significance occurs through what Cresswell terms the 'production of mobilities' (see the following box).

Key idea 2.1: THE PRODUCTION OF MOBILITIES

Tim Cresswell's writings on place, power and modernity have provided some of the most influential contributions to the 'new mobilities' lexicon. Cresswell appears frustrated by the clear lack of attention given to mobilities by geographers and social scientists as a concept that is just as important as themes such as place, space and society. Cresswell's strongest critique revolves around the relationship between mobility, meaning and power. The humanist tradition within geography is clearly evident in Cresswell's argument when he contends that without an appreciation for the symbolic role and meaning of mobility, we are left with simply movement (see Canzler *et al.* 2008 for a different reading of this division). To Cresswell, meaning is the vital ingredient missing from this sort of study. In *On the Move* (2006a: 2) he explains:

> I want to make an analytical distinction here between movement and mobility. For the purposes of my argument let us say that *movement* can be thought of as abstracted mobility (mobility abstracted from contexts of power).

If we follow Cresswell's argument, mobility has been investigated as 'the general fact of displacement before the type, strategies and social implications of that movement are considered' (Cresswell 2006a: 3). By drawing on efforts to understand space in terms of its social and experiential dimensions, Cresswell aligns mobility as something akin to the idea of place: '[M]obility is the dynamic equivalent of *place*' (Cresswell 2006: 3). From this point of view, without meaning we are left with something rather superficial. We have simply movement, that is it. This is obviously problematic in two main ways. First and foremost, 'Movement is rarely just movement; it carries with it the burden of meaning' (2006: 7). Thus, to ignore the way movement is entangled in all sorts of social significance is to simplify and strip out the complexity of reality as well as the importance of those meanings. Second, we must question approaches that do treat mobility as otherwise. To abstract mobility to movement (mobility without meaning) often has distinctly political consequences.

Further reading

(Cresswell 1993, 1997, 2001)

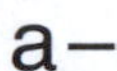

Figure 2.1 Abstracted mobility

Starting right at the most basic principle of Cresswell's thesis, we can consider a diagram I have repeated in Figure 2.1.

This schematic summarizes some of the more basic ways mobility has been understood in academic thought. By positing mobility as simply a movement from a to b, the diagram is one way to visualize the reduction of mobility to an action of simply getting from one place to another. A and b – the departure and arrival points – are fleshed out in our diagram as letters, yet the line in-between remains just that: simple, misunderstood and negligent of a much more complicated reality. As Cresswell puts it, 'the bare fact of movement ... is rarely just about getting from A to B.' Cresswell wants research to look at the movement represented by the line that connects the two points together. Regardless of the line's supposed immateriality it is 'both meaningful and laden with power' (2006: 9).

If we tease out Cresswell's argument a bit further, there is something in the space between the two points, something about the context of mobility, that makes a vital difference. In other words, Cresswell sets out to ask: how are the kinds of meanings attributed to any sort of mobility entirely contextual? The meanings I ascribed to my examples of meaningful mobilities earlier are partly a product of 'my' own placing. It does not mean that I cannot escape the position of who I am, but they surely shape my interpretation. Quite different meanings could and would be ascribed by anyone else. Mobility has no pre-existing significance in and of itself. Mobility does not implicitly mean one thing or another. Mobility is not essentially good or bad, suggestive of a vacation or commuting to work. Rather mobility is *given* or inscribed with meaning. Furthermore, the way it is given meaning is dependent upon the context in which it occurs and who decides upon the significance it is given. Taking this approach builds up a picture of much more than a dotted line across a blank page, but mobilities travelling over and through the complex terrain and topology of social spaces (Figure 2.2):

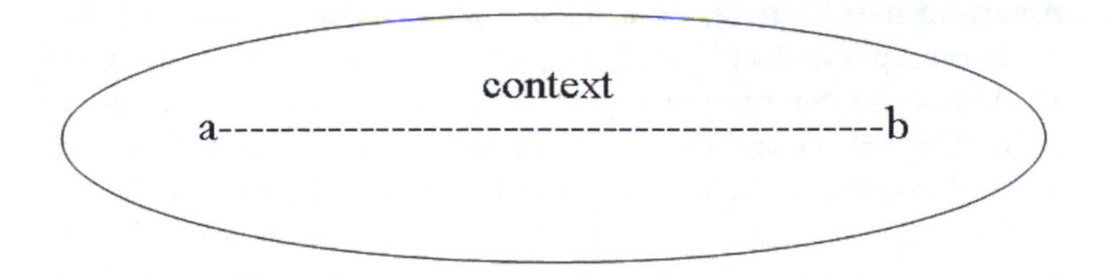

Figure 2.2 Mobility in context

In this light, the meanings given to the mobilities could be read quite differently. They could be given much more threatening inscriptions. To a landowner a walk in a wood might be viewed as an intrusion or a threat to their private property. A walk down a street, to someone walking the other way, might be perceived as a threat to their safety. A stroll off the footpath could be interpreted as a menacing act to the integrity of the surrounding environmental habitat by a local environmental interest group. In other words, mobility gains and is attributed meaning by those who interpret and make sense of it.

Even while mobility has no pre-existent meaning, certain places, cultures and societies can give particular kinds of mobility particular kinds of meaning. These meanings might even cross over cultures. Consider how a simple vector of mobility has a remarkably similar symbolism over the past 2,000 years. Garrett Soden's work on falling reminds us how 'Human culture is filled with stories that equate falling with failure.' From Icarus to Lucifer to leaders *falling from power* even '*falling in love* implies a loss of control, as does to *fall asleep* or to *fall under a spell*' (Soden 2003: 15). Downwards mobility is often attributed with negative significance. As Anne Game (2000) shows, too, falling is associated with a kind of passivity and the loss of self-determination over one's fate.

Vertical social mobility can of course be related to geographical forms of movement. Moving away for a new job or another opportunity, for example, can often mean moving up the social ladder. Wilbur Zelinsky (1973) attributed these sorts of associations with the social and geographical mobility of the citizen of the United States: 'The

American never arrives; he is always on his way' (1973: 58). The American's physical and social progress were inseparable. As if one of Eli Whitney's interchangeable parts (Hounshell 1984), the American is, 'highly versatile and movable, and eager to insert himself into the locus of maximum advantage to himself and thus, of course, to the system' (1973: 59). Staying put in today's society may be viewed to be quite negative, thus acting as a barrier to ascendance up the social ladder. As David Morley (2000: 202) has suggested, mobility 'is increasingly seen as a social good and immobility increasingly acquires, by contrast, the connotation of defeat, of failure and of being left behind'. Envisage the plethora of advertising surrounding all sorts of computing technologies. Laptops are advertised for their portability – most often their mobility (Mackenzie 2006). The laptop I write this book on is known as an *ultramobile* model of computer and its advertising promotes all sorts of social aspirations.

Again, one's social situation or context may not necessarily see another's upward and geographical mobility as something particularly good or positive. For instance, the rural communities experiencing inward or outward migration may not see rising house prices or, conversely, unfilled jobs and declining services as particularly good (Cloke *et al.* 1995; Cloke *et al.* 1997; Milbourne 2007). This chapter takes this context and contingency seriously, exploring how mobilities-as-meaningful are constructed through different contexts and positions.

We must consider further that these meanings, while ephemeral and fragmentary, can achieve considerable presence. The brush we tar mobilities with can leave quite a permanent stain. And the meanings given to mobilities make a difference. In fact, they can make a big difference. They can shape social relationships, and they might alter the way we think about and act towards them. For some people, labels have an intrusive and permanent presence which will simply not go away (on the stigma given to gypsy travellers see Halfacree 1996; Holloway 2003, 2005). But just as our concern for mobility should consist of looking at the world, we must also consider how these meanings are pervasive within the academic study of mobility itself. They may mirror, reflect or alter those of wider society. As David Morley usefully argues, 'The very vocabulary in which we discuss questions of mobility is, as we have seen, inevitably value-laden' (Morley 2000: 41).

Dwell for a moment on a definition of mobility by transport geographers. In the introduction to their classic text *Modern Transport Geography* (Hoyle and Knowles 1998) the authors state 'in all societies, environments and economies the movement of goods and people – as well as capital and ideas – is a necessary element in functional and developmental terms' (1998: 4). Clearly at the time Knowles and Hoyles' idea of transport mobility did not encompass the social and cultural senses of mobility as a vital ingredient (although Knowles' later work does, see Chapter 5). But to weed out mobility from purely functional and developmental terms is not to say that meaning and social significance do not exist, just that it is not always looked for in that way. The authors have made their own value judgement of what the study of mobility should be, even if it precludes mobility's social significance. We must, therefore, ask where was this judgement made and from what position?

From transport geography, Hoyle and Knowles' (1998) writing reflected and emphasized the utility of geography as a subject involved in the analysis and planning of transportation infrastructures and policy. The specific disciplinary context or space of transport geography played an important role as the background from which Knowles and Hoyle made sense of mobility. In short, the key is *context*. Whether 'out there', or within the confines of academia, context makes a difference to the ways mobilities are given meaning and understood. Having said that, we should be also careful in the ways we imagine context. Geographer Marcus Doel (1999) argues that spaces may be imagined as quite fixed and discrete containers. As a note of caution then, mobilities must be seen as involved in the making and remaking of spaces and contextual backgrounds. To understand how they achieve significance is to make sense of their impact upon foundations of shifting sand - their implications for disciplinary debates and assumptions moved-on. If we turn to Knowles (2006) and other more recent writings on mobility, their ideas have developed in ways that embrace the 'new mobilities paradigm' as a more culturally nuanced perspective to examine transportation today (Pirie 2003).

FIGURES AND METAPHORS OF MOBILITY

Mobility is often addressed through two fairly distinct viewpoints (Cresswell 1993; 2001; 2006; Kaplan 1996; Urry 2000). In Lisa

Malkki's (1992) examination of transnational identities, she outlines what is described as both a 'sedentary metaphysics' – the propensity to see the world in fixed and bounded ways, or, its antithesis: a 'nomadic metaphysics' – a way of seeing that takes movement as its starting point. While these categories of thinking about mobility are useful, there is a tendency to lump together some really diverse approaches, the nuances of which often become obscured in an effort to highlight their similarities while overlooking their differences.

It is my intention to try to unpack these two sorts of thought in greater detail and to add flesh to several other figures or metaphors that are deployed to comprehend mobility within these styles of thought. The way I wish to do this has several dimensions. First, we will take sedentarism and nomadism in turn while tackling several of the figures and metaphors of mobility that exceed these initially simpler divisions. Two figures underpin these bodies of work.

The first more associated with sedentarism fetishizes the figure of the network node. An immobile point, the node's primacy means that mobility is merely its servant – a means to get from one node to another. The node, we will see, is meaningful while the line of movement is merely the method to get across meaningless space and to the sanctity of the nodal place. Furthermore, the node's borders – its circular boundaries – serve as impermeable barriers. The inviolable place is kept safe and secure from outsiders. Nomadism, on the other hand, takes the line – which stands for mobility – as its starting point. Under nomadism, it is the line that is privileged and not the node. Flow dominates the points and pauses that emerge as places in between and on the way to going somewhere.

Sedentarism: nodes, networks and places

Beginning with some of the first and most influential work to examine mobility, cultural geographic studies of early modern societies took issue with certain formulations of nomadic peoples. Instead of romanticizing their apparent wanderlust, the life of the nomad was seen to be rather more sedentary, immobile and, therefore, more meaningful. Take Berkeley scholar Carl Sauer who looked at the origins of early modern agricultural societies. For Sauer (1952), apparently the nomadic groups moved as little as possible, moving on for the basic needs of 'food, water, fuel and shelter'. Sauer went on to explain how 'Mobility as a dominant character goes with specialized hunting economics or with life in meagre environments' (Sauer 1952: 22). Mobility was an activity more appropriate for hunter-gatherer societies or those living in environments that forced the continual search for food. Mobility was induced or necessitated as an exception. Mobility was a threat to all of the 'good things' found in sedentary living. Sauer wrote of 'vagrant bands, endlessly and unhappily drifting about', which was to equate mobility with indecision, perpetuity, and emotion.

This sort of approach appears to laden mobility up with the heavy baggage of economic rationality. Sauer draws upon early principles of economy and least effort suggesting that 'our kind has always aimed at minimizing' the costs and exertion of movement. In this light, mobility and migration were used infrequently and only when necessary. Relocation was something forced through seasonal changes and when another location required far less effort to subsist in. The 'hearth' and the 'home' acted as gravimetric attractors. The hearth worked as a centre around which family life orbited and goods were collected. Hearths were usually patterned in such a way that showed their deliberate positioning at points 'of least transport'. In saving energy and effort, they were also relocated as 'infrequently as necessary' (Sauer 1952: 12).

Sauer's interpretation was not necessarily unique. Although working from quite a different perspective, a similar emphasis may be found in the research of French geographer Paul Vidal de la Blache. Like Sauer, Vidal de la Blache's (1965) humanism was at pains to explain how nomadic groups were geographically tied down in one way or another, saturated by their physical milieu. For starters, they were always seen to 'bear a definite relation to a given amount of space'. It was seemingly inconceivable that a group of people could exist without the need of a particular space or a place to return to. He wrote how it was 'within neither reason nor experience that a people should exist without roots, that is, without a domain in which to carry on its life activities, one which will ensure and provide for its existence' (Vidal de la Blache 1965: 52).

Sauer's communities were a step-away from nature by their insistence upon stability, fixity and the pull of the hearth. On the other hand, Vidal de la Blache simplifies the nomad's way of life, whose motivations and actions are determined by the environment and the landscape around them. Geographical regions offered 'natural facilities for getting about'. These were the sort of regions where erosion had levelled off the mountainous terrain to manufacture a smoother surface mantle ideal for movement upon it. The mobilities of the nomads made them much closer to the natural and the animal life they had left behind on the evolutionary scale. Their movements were described 'like herds of animals roaming over the steppes, or great flocks of birds swooping down upon stretches of water' (1965: 361).

In these descriptions, the French scholar reduced the human to a sort of animal-like barbarian who moved in 'hordes' (Vidal de la Blache 1965: 361). Mobility was not meaningful but instinctive, gestural and determinable. The nomadic movements of Turks, Mongols, Magyars, Bulgars and Huns are described as 'periodic spectres, which, with horses and carts, came forth from the world of the steppes as from their natural habitat'. They are rendered entirely determinable and predictable because they appear to act in just the same way as natural phenomena might do. As it was put, 'In the swiftness with which they took shape and the definiteness of the trajectories within which they appear to have been held they resemble meteorological phenomena of which science can determine the origins and follow the paths' (Vidal de la Blache 1965: 368–369). Invariably, the reductive simplification of certain kinds of human mobilities to nature were levelled at societies other than Vidal de la Blache's own, where a 'Hindu ant hill' was 'kept constantly in motion by requirements of commerce or religion, as well as by mere Wanderlust' (Vidal de la Blache 1965: 384). Simple rules could determine eventual movements so that when the 'the hive is too full, a swarm leaves it' (1965: 70).

We are clearly seeing interpretations which are products of a definite time and place – the time and place of a developing regional geography and other ethnocentric interpretations of societies and their surrounding landscapes (see for instance Robert Park's urban sociology). The nomad and suppositions of nomadism are treated with almost disdain. It is 'inconceivable' that people would live in such a way without attachment or commitment to a particular and stable place. Mobility is identified as especially primitive or primeval. And even while simplifying metaphors are deployed to represent them, Sauer and Vidal de la Blache work hard to give early societies rootedness and a meaningful engagement with places.

Mobile method 2.1: EXCAVATING AND RECONSTRUCTING MOBILITIES

Typically the fields of archaeology and anthropology have shown how the mobilities of early settlements, while 'difficult to document', may be identified by their traces (Marshall 2006). Evidence such as tools found far away from the source of the raw materials required to make them, can provide an indication of residential, logistical, territorial mobility and trade. Certain kinds of tools may even gesture towards the frequencies of these mobilities. Strategies of mobility reconstruction are prone to high degrees of uncertainty, given the many other variables associated with the spread of tools. Thus, different kinds of objects can be used to trace early societal movements such as rubbish and waste. The abundance and diversity of rubble, animal bones and other matter provide indicators of residential movement (Kelly 1992). Kelly explains how, 'instead of simply throwing or sweeping trash off to the side (as mobile foreigners do) . . . sedentary Baswara used secondary trash dumps located farther from houses than those in camps of residentially mobile groups' (Kelly 1992: 56).

Other approaches have taken the form and structure of traces such as buildings and monuments to imagine mobility patterns and paths from plans and diagrams reconstructed from the material remnants of a site. Neil Turnbull and other archaeologists have argued that taking this 'building perspective' sees such movement patterns to be reflective of a societal and physical structure, rather than making of it. Questions by writers such as Chris Tilley, 'how people move through monuments, what they see from different points, how the physical experience of the monument affects its perception' (Johnson cited in Turnbull 2002: 132) offer more complex than deterministic interpretations. If the activities and movements of people shape and alter these physical structures a further recursive interaction takes place, reshaping the 'movement of people to particular places, to limit posture and what could be open to view' and hence 'the ways in which a place could be experienced and read' (Thomas cited in Turnbull 2002: 134).

In more immediate contexts, mobilities leave footprints that researchers can follow. Archival and historical documents may be consulted for simple descriptions of mobility, to much more detailed, qualitative experiences captured in oral histories or the moving image. Historical methods have allowed researchers to reconstruct experiences of mobility from company records and oral testimonies. Peter Merriman's (2007) work on the M1 motorway is a particularly good example as the mobile culture of the road is brought to life (see also Robertson 2007).

In practice

- Mobilities may be reconstructed by finding their traces from the activities and practices of societies
- Materials, remnants of buildings, historical documents, texts, photos and objects are all useful objects of analysis
- These interpretations can require situating within their contextual cultures of activity, bodily perceptions and experiences

Further reading

(Kelly 1992; Turnbull 2002; Marshall 2006)

Laws

In other approaches, the meaning and significance of mobility is addressed in an altogether different manner. A host of laws and allegories are investigated for the way they were intended to better describe and *explain* movement. Despite these marked differences with those above, their treatment of mobility shares several commonalities with the regional geographic work explored. While the network diagram of nodes and lines was more metaphorical in the approaches we have seen, in these investigations they are used directly as representational models and norms through which the reality of mobility is understood as a function of transportation, migration and industry.

For efforts at refining laws or single principles that govern *why* people and things move or become mobile, take Abler, Adams and Gould's (1971) classic *Spatial Organization* from geography. The authors want to study why things move in order to uncover and understand 'movement laws'. They do so because they 'want to predict and control social and natural events' (1971: 238). These explanations could be based upon very simple factors. Ullman defined three. The first was 'complimentarity' – how differentiation of space could promote interaction (Ullman 1957). The second he called intervening complimentarity, or the opportunities between two regions or places. And finally by 'transferability' Ullman measured space or distance in terms of the cost and time of transport.

At first glance, the above investigation appears to bleed many of the social dimensions out of the mobilities they study towards a far more functional analysis. And yet, they are undertaken in order to tackle various social problems. Significance is found within the kinds of relations, associations and patterns such studies uncovered, from the access of services to the location of industry. But it does not mean that value-laden judgements, metaphors and figures are not used to make sense of mobility.

One of the most famous investigations of the laws of movement was developed within the subfield of social physics in the 1960s. Attracted by the laws of movement, matter and energy within the physical sciences, social physicist James Stewart (1950) pondered how human mobilities might behave along the same principles as physical materials. George Zipf (1949) directly appropriated the principle of the 'conservation of energy' into a pseudo-psychological and ecological theory. In his hybridization of this law Zipf applied it to human activity, Zipf argued that 'every individual's movement, of whatever sort, will always . . . tend to be governed by one single primary principle which, for the want of a better term, we shall call the *Principle of Least Effort*' (Zipf 1949: 1).

Zipf's approach influenced and resembled several other ways of addressing mobility. A dominant approach was to see it as an activity used to simply overcome the *tyranny* of distance. Mobility was taken in instrumental terms as a means to get from one place to another as quickly as possible. As a functional tool to overcome distance (as we saw earlier in Hoyle and Knowles' formulation) mobility was to be endured. Walter Christaller's famous traffic principle offered up a 'movement law' that saw the arrangement of central places as most efficient when the routes between them could be established 'as straightly and as cheaply as possible'. Advocating what we will see later as a form of network bypass (Graham and Marvin 2001), central places could be connected together with speedier mobility linkages that would see 'the more unimportant places . . . left aside'. Central places should be lined up on traffic routes in the most efficient manner possible (Christaller cited in Lloyd and Dicken 1977: 44–45).

At the time of this research economic theorists such as Walter Isard (1956) were influential in the social sciences (Chang 2004), following Stouffer's theories of mobility and opportunities (Stouffer 1940: 846). Mobility underwent abstraction as it was lifted off into a rational movement equation. Simple variables and inputs were plugged into formulae that could be repeated over and over again. By adding in physical characteristics, the distance that a unit weight would travel, how it would correspond to an exertion of effort, and a factor of services required to overcome this resistance, or what Isard called friction, the academic could understand and *explain* mobilities in order to *predict* future possibilities.

Again consider the values embodied in mobility here. Isard's aims are as transparent as can be. Space and the friction of space quantified in terms of cost and effort must be overcome. Christaller posed the similar issue of how space or *economic distance* should be 'determined by the costs of freight, insurance, and storage; time and loss of weight or space in transit; and, as regards passenger travel, the cost of transportation, the time required and the discomfort of travel' (Christaller 1966: 22). Time and, therefore, movement equalled capital and it was imperative to minimize this cost. The complex characteristics of mobility and its contexts are simply taken as variables; quantifiable entities to be swapped into an equation.

The second theme within this sort of work was the figuration of mobile people not as input variables but as imaginings of atomized individuals listening not to the pain of their limbs, but to generalized gravimetric forces radiated by masses of different size. The gravity of a metaphorical mass explained why people would prefer to travel particular distances and to particular places. One of the earliest and most famous investigations in this area was conducted by Ravenstein (1889) who attempted to formulate a theory of migration. Adopting several fluidic analogies of streams and rivers, circulations and eddies, Ravenstein argued that mobility was directly proportional to distance. He presented the idea that migration movement was dependent upon the distance travelled, thus, 'current lost in strength [is] in proportion to its distance from the source of supply' (Ravenstein 1889: 286–287).

In this formulation Ravenstein was able to reject the notion that people had moved about the 'earth's surface like a sheet of oil, spreading slowly and evenly in all directions' (Vidal de Blache 1965: 71) but rather envisaged certain centres of undeveloped resources, or over-population elsewhere which would provide pulls and pushes on migration movement. Even as it first appears that Ravenstein is completely physicalizing movement and the causes of mobility, he does suggest that 'the desire inherent in most men to "better" themselves in material respects' (Ravenstein 1889: 286) is persistently the most powerful driver of movement. Both causes are conflated. It is not that the call of surplus land or the need for labour provide qualitative urges to move across vast distance. He rather envisages interlinking waterfalls and cascades of mobilities and opportunities creating both physical and qualitative momentum, pulling travellers from province to province.

Academics influenced by these ideas did not assume a literal gravity field generated by cities but a metaphorical one experienced in a social way (Hua and Porell 1979). Gravity in these studies meant influence, it meant opportunity, and it meant desire (Olsson 1965). However, we should not mistake desire for an emotional kind of craving, lust or impulse (Olsson 1991). A rather less feeling and more thinking subject was imagined: 'rational economic man' who would respond to external forces with calculation and expediency.

Analogies such as these placed a direct superimposition of 'molecule' upon 'man'. The primacy of the figural nodal points we have been thinking about were dropped onto the individual. Mobile people had *mass*. Individual people behaved like molecules and aggregations of mobile bodies would act like a collection of molecules with increased molecular weight, 'The greater number collected in a given space, the greater is the attractive force that is there exerted' (Carey cited in Dicken and Lloyd 1977: 150). Ideas such as these allowed academics like Dicken and Lloyd to view movement and interaction 'as a variant on the general physical law of *gravity*' (Stewart and Warntz 1959; Lloyd and Dicken 1977: 96).

The way this was manifested mixed physical metaphors with economic theory. These approaches generated what could be imagined as a particularly uneven snooker or pool table. The individual is taken as a mobile ball unable to conjure any internal agency to drive their own movement. Rather the physical and economic structure of the table determines the movement of the balls. Distance and opportunities create peaks and troughs in the table of gravimetric potential of attraction and repulsion that push, pull, yank and tug at the balls with the only resistance coming from the surface of the table and the distance to be travelled. While obviously problematic, this approach saw remarkable utility in the prediction and understanding of human and non-human mobilities. It saw application to understand the movement of consumers to central places, of goods to consumers and of agricultural products to central markets (Dicken and Lloyd 1977: 65).

These approaches found popularity within a certain societal and academic context. In the context of the Cold War, as well as post-war demands for reconstruction and urban renewal, the laws of mobility were sought in an academic charge towards scientific legitimacy, leaving mobility as something that *needed* to be explained and predicted by laws (see the box below).

Case study 2.1: SPATIAL SCIENCE AND THE COLD WAR

As Trevor Barnes documents, the imperatives of post-war development and Cold-War planning meant the increasing relevance of geography and, specifically, what geographers could say about mobility and transport. In the context of pressures to chase grants that would lead to the planning of public mobility infrastructures, similar values were attained. As policy makers demanded scientists and experts find the answers to traverse distances in the United States, this caused increasing demands upon the understanding of mobility itself.

In the United States William Garrison's 'remarkable' work achieved high profile with definite utility for public funding. His expertise was sought for the planning of highways, such as the Washington State Highway Commission funding for a potential highway system around Seattle. Elsewhere, Garrison's work was used in the context of civil defence evacuation schemes in order to plan optimal and efficient movements out of cities by car and ferry. It was the necessity to manage and shape the movement of people which saw increasing attempts by geographers to respond to this call, and provide the means for policy to do this.

As Barnes and Farish note, what also distinguished this sort of analysis was the match-up between the knowledge and analytical tools geographers employed and the reality – the real world mobilities – they attempted to understand. In Garrison's case, his team's development of 'lines of flight and the systemic grids of emergency response' for public evacuation in their work for Operation Rideout 'represented a rational ideal' and furthermore, a 'refined unreality of abstraction' (Barnes and Farish, 2006: 819). Such 'refined abstractions' took many forms. The sorts of metaphorical of molecules, atoms and flows we have seen were visualized in modelling diagrams. Indeed the very philosophy of models reinforced the epistemological and ontological trick of spatial science which attempted to eliminate complexity in order to enable comprehension.

Further reading

(Farish 2003; Barnes and Farish 2006)

Taken from engineers and physicists' visualizations of electrical circuit diagrams models of movement simplified mobilities down to the sort of line I showed at the beginning of the chapter. The individual moving between two points in space forms the *route* or the line, and then the points at which the mobilities start, finish, or are focused, form what were described as *nodes*. Other ways of visualizing these relationships drew upon physicalist metaphors such as flows and rivers. Peter Haggett (1965) used river systems and their hierarchical system of networked tributaries to make sense of the order of mobilities. In wider fields such as architecture, the shape and form of actual river systems provided one of the greatest inspirations to modernist architect Le Corbusier in his designs of urban master planning. 'What an invitation to meditation, what a reminder of the fundamental truths of our earth!,' Corbusier exclaimed on a flight over the Amazon (Pinder 2004: 82).

In their own terms, these approaches marked major breakthroughs in the analysis of mobility. Indeed, their legacy in more sophisticated investigations within fields such as transportation geography, have proved incredibly useful for the detailed and descriptive studies of mobility systems, routes and patterns (see Chapter 5). And yet, even though these approaches towards mobility were focused upon law making and predictive assertions of mobile behaviour, this did not necessarily mean that what we might consider to be more qualitative dimensions of movement were not considered or imagined. Gunnar Olsson suggests that 'not only might the questions of "how" and "where" be answered, but hints might also be given as to why people obey the behavioural "laws" mentioned in this and other studies' (Olsson 1965: 73). While laws could be developed from the analysis of people's movement patterns – the surface – issues of *why* could only be posed if one was to understand people's 'non-rational' behaviour and motivations.

Lloyd and Dicken (1977) make it clear once more that the simplistic, reductionist principles of an *infinite mobile* is far from perfect and may well describe patterns of mobility only in general terms. They write,

> How, for example, do our watchmakers get to hear about the higher wages? What happens when a competitor in the next state raises his wages to watchmakers? Do they pack up and move again? What do their wives and children think about all this mobility? What about the low-paid watchmaker who likes it in Peoria, Illinois, because he grew up there?
>
> (1977: 206)

As they move on to state, the reality is rather more complex and tricky to predict.

It was this kind of complexity that time geographer Torsten Hägerstrand (1982) sought to uncover through an approach that took the atoms, points, lines and paths to a degree far beyond anything from spatial science. Hägerstrand unpacked the lines of mobility in order to explore the individual biographies of mobile subjects far more seriously. His technique worked to atomize subjects into individual particles with *projects* determined by various individual time–space constraints or potentialities (see below).

Key idea 2.2: MOVEMENT, PATHS AND SPACE–TIME ROUTINES

Within geography an incredibly influential force for the study of mobility was Swedish geographer Torsten Hägerstrand. The foundation of his time geography was a conceptualization of time–space: any movement in space was also a movement in time. Focusing upon the routines and paths of individuals as 'biographical projects' Hägerstrand sought to understand how individuals' movement patterns and activities did not happen in a vacuum, but rather how they interacted with the *milieu* of spaces and places. Moving about in space and time, Hägerstrand's approach was notational in that it sought to trace out the characteristic shapes and patterns of people's routines.

The spaces and times through which people moved created what he defined as 'constraints' upon one's capacity for mobility and to undertake their projects. What he imagined as time–space prisms were the physical/spatial and temporal walls within which people could act. Thus, a particular trip or what he called a project would take a certain amount of time, which would be constrained by the volume of time–space available on a daily basis.

These were understood as fields, fields of movement radiating out from a centre which was usually their residence. Thus, 'places of work, shops, places of recreation, residences of intimate friends and other similar locales serving as nodal points' (Hagerstrand in Gregory 1985: 306) were the stations through which time–space routinization occurred.

Further reading

(Hägerstrand 1982; Giddens 1985; Gregory 1985; Hägerstrand 1985)

As represented below, Hägerstrand produced memorable time–space diagrams which visualized movement through space and into the conduct of everyday life.

Like the spatial scientists before him Hägerstrand repeats the imagination of the mobile individual as physical atom, reinforcing the equivalence quantitative geography imposed upon humans and non-humans. In arguing that geographers should take a 'world on the move, a world of incessant permutations' (Hägerstrand cited in Crang 2001: 192) much more seriously, he saw how such life-paths could be easily applied to 'all aspects of biology, from plants to animals to men' (Hägerstrand in Gregory 1985: 311). And yet, Hägerstrand did not mistake his representational models as perfect stand-ins for the biographical projects they represented. He noted that the paths he traced 'in the time-geographic notation seems to represent nothing more than a point on the move'. It was

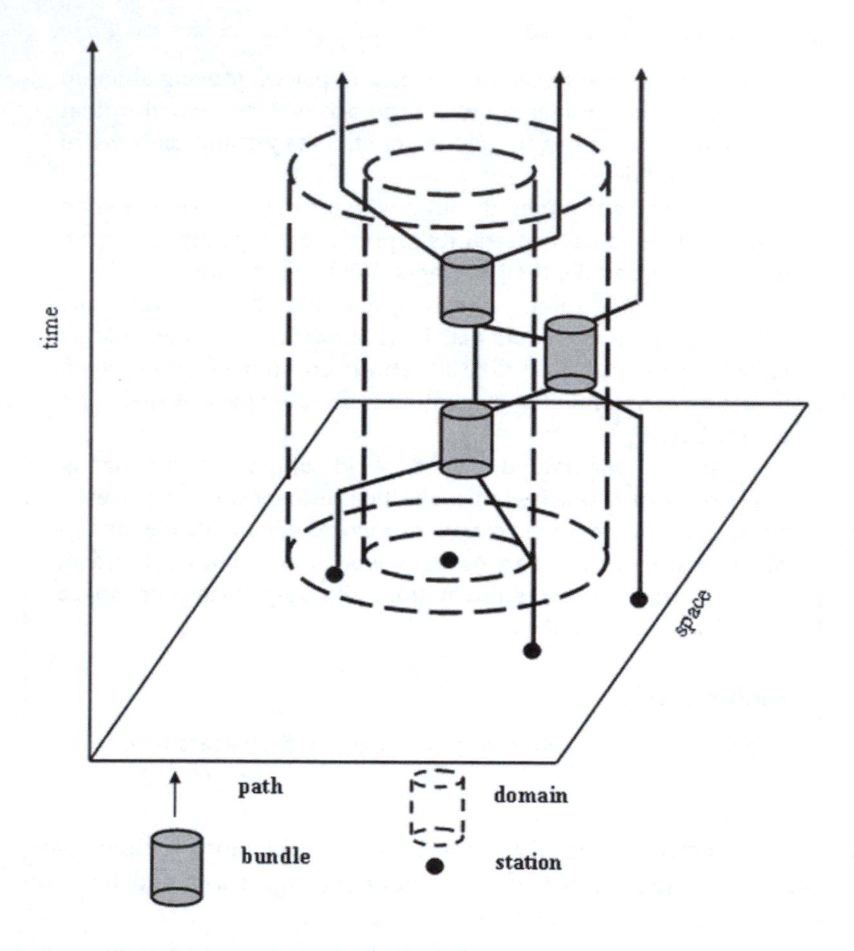

Figure 2.3 Time–space path visualization

Source: After Hägerstrand (1967, orig. fig. 1.4.4)

important that they 'should not lead us to forget that at its tip – as it were – in the persistent present stands a living body subject, endowed with memories, feelings, knowledge, imagination and goals'. Thus, although these variables were far 'too rich for any conceivable kind of symbolic representation', they were, in fact, 'decisive for the direction of paths'. People, he went on, 'are not paths, *but they cannot avoid drawing them in space–time*' (cited in Gregory 1985: 324). It is an awareness of this depth of experience that we can now turn to.

Mobile method 2.2: FOLLOW THE THING

As emphasized so far, research on mobility is concerned with objects as well as mobile people. Mobility research has posed methodological questions for researchers interested in the paths, circuits and social fields objects travel through (Shanks and Tilley 1993; Ingold 2007b). Geographer Ian Cook's studies of food networks advocates the methodological practice of following-the-thing. Cook's work follows an object such as papaya as it is passed through a series of spaces and chains that see the object transformed. Cook's primary interest in this processing is not necessarily the transformation of the papaya but the experiences of the Jamaican labour forces who conduct these processes.

Following the movement of a papaya as it is picked by Jamaican farm workers, as it is carried, transported, washed, weighed, graded, tripped, wrapped and packed, highlights a host of labour practices that take place in concerning social and environmental settings which the researcher may in turn see, touch and experience. Humanizing what appear as disembodied food networks thus highlights the very human and de-humanizing labour conditions workers must endure, the treatment of animals and other commodity chains (Goss 1999).

In practice

- Ethnographic approaches towards objects may mean physically following their movements through different social and geographical settings
- The recording of direct observations and recollections of these settings can provide a useful snapshot of the processes and conditions through which objects such as food are harvested, prepared and moved

Further reading

(Cook *et al.* 1998; Cook 2004, 2006)

Places and placelessness

As a student of Sauer, Yi-Fu Tuan was an instrumental force in the humanistic geographic approach which departed from and greatly criticized the perspectives of a positivist view from nowhere. Tuan's writings explored how people made their lives significant through experience. Interested in how places served as home spaces (Tuan 1977) imbued with meaning, just like Sauer Tuan found it difficult to see past the need for people to form roots and attachment to places. From this perspective, mobility place are 'antithetical'. Whereas, 'movement takes time and occurs in space', place is posited as 'a break or pause in movement – the pause that allows a location to become a centre of meaning with space organised around it' (Tuan 1978: 14). If we could visualize this, Tuan's focus is on the points, not abstract dots but points as foci of meaning and significance with space lying in between them.

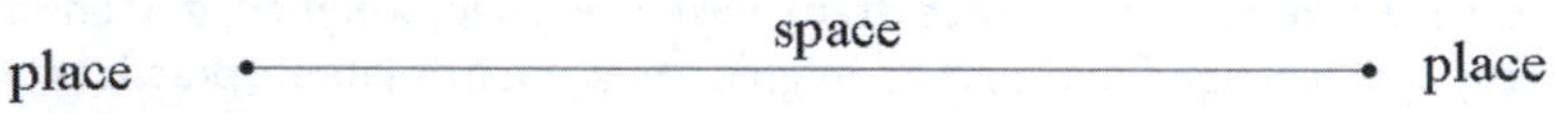

Figure 2.4 A schematic of Tuan's theory of place

Unlike the spatial scientists who visualized these points or nodes as physical and economic attractors, Tuan gave these points both context and significance. Places were *centres of meaning* or points around which we organize our social and inherently meaningful lives.

A popular counter figure to be found within Tuan's work was once more that of the nomad (Tuan 1974). Human mobility was taken as the direct aggressor to points or places. 'To be always on the move' was understood as a way to depart from or 'to lose place'. One on the move was, therefore, out of place or 'placeless' holding only superficial 'scenes and images'. Tuan's contemporary, Edward Relph, decried the negative imposition of roads, railways, airports and spaces for movement itself. Relph suggested that by 'making possible the mass movement of people with all their fashions and habits', these spaces had led to the 'spread of placelessness' (Relph 1976: 90).

It is just as inconceivable for Tuan and Relph that nomadic lifestyles of pure and unattached mobility should occur. When they did, they were cast in the shadow of quite morally charged ideologies that saw their movement as threatening and negative. Similar writers in this field such as landscape author John Brinckerhoff Jackson (Jackson 1984) demonstrated a preoccupation with the *sanctity* of place. From this perspective, it is place, or the 'permanent position . . . that gives us our identity' and furthermore for Jackson, 'whatever is temporary or short-lived or moveable is not to be encouraged' (Jackson 1986: 194). He wrote about how he was 'confused by the temporary spaces I see: the drive-in, fast food establishments that are torn down after a year, the fields planted to corn and then to Soya beans and then subdivided; the trailer communities that vanish when vacation time is over' (1986: 155) and how could one find home, solace and significance in these sorts of place? On the same kinds of landscape, Jean Baudrillard's (1988) infamous *America* wrought a breathless journey of automobility across the deserts of the Southern states, wherein the heat, the mobility and the banality of the landscape combined to produce an 'evaporation' or 'extermination of meaning' (8–9).

These sorts of perspectives are not limited to academia. Similar points of view have often had very problematic consequences for those on the other end of them. Cresswell voices some of these concerns by expanding the metaphysics of sedentarism beyond the *internal* debates of geographers, sociologists, cultural theorists and others. Cresswell argues that 'The view of the world that attaches negative moral and ideological codings to mobility extends well beyond the ivory tower to pervade thought and practice in multiple domains of social and cultural life' (2006: 38). While we will look at the politics of this kind of sedentarism through processes of exclusion in the following chapter, the case study box below is devoted to the problematic implications of just this sort of thought. Xenophobic discrimination in Southern Africa is rendered all the more convoluted and paradoxical, given that reactionary sedentarist thinking is born from the nomadic ideology of neoliberal capitalism.

Case study 2.2: CITIZENSHIP AND SEDENTARISM IN SOUTHERN AFRICA

Francis Nyamnjoh's (2006) *Insiders and Outsiders* departs from Western-centric examples of the devastating implications of sedentary thought through his detailed case studies of xenophobic discrimination in Southern Africa. Investigating the way immigrant Africans are imagined and treated in South Africa reveals a complex process of discriminatory treatment justified by several metaphors and figures of mobility.

Various investigations of this issue show how sedentarism has shaded certain migrants with racial and pathologized stereotypes, conjuring up ideas of 'nightmare citizens' whose 'rootlessness' sucks and siphons the morale and economic value from South Africa herself (Comaroff and Comaroff 2002: 789). Nyamnjoh's study focuses in part upon the construction of the *Makwerekwere* – the name given to mobile immigrants of African origin as 'locals' respond to the inequities of the social and material transformations of South Africa led by the rise of neoliberal capitalism, intensifying market competition, dividing labour and the rendering of increasingly porous national politics and economies as Comaroff and Comaroff explain (2002: 797).

Denying these people an apparently intelligible language, South African insiders claim only to understand a 'barbaric' form of 'stuttering'. Aligning the migrant population with dehumanizing and animalistic signatures, South African stereotyping qualify the *Makwerekwere* 'as the "homo caudatus", "tail-men", "cavemen", "primitives", "savages", "barbarians" or "Hottentots" of modern times' (Nyamnjoh 2006: 39). Through a hierarchy of race and skin pigment the *Makwerekwere* are believed to be the 'darkest of the dark-skinned' and, therefore, to be the less enlightened. Incoming groups are regularly perceived as alien, illegal and only there to steal the jobs and opportunities that would have gone to local people.

The *Makwerekwere* are believed to bring the problems of neighbouring countries with them: civil war, genocide, dictatorships, corruption and crime are all candidates. As Nyamnjoh puts it, 'Suddenly having to face an influx of primitive darkness in the urban spaces of the new South Africa could be quite disturbing, indeed a nightmare from the past' (40). The mobility of the *Makwerekwere* is interpreted as organic and plant like using up the local resources, spreading like a weed, and strangling the opportunities for local people. Interpreted as a physical and moral contagion, they are believed to have 'come to poison South Africa with strange diseases such as AIDS. Their women readily hold on to the arms of impressionable South African men, whom they dazzle with 'sugar-coated kisses that . . . [are] sure to destroy any man' (Nyamnjoh 2006: 44).

Further reading

(Cresswell 1996; Comaroff 2002; Nyamnjoh 2006)

Nomadism: lines of flight

In the above discussion several figures of mobility have been tied down in various ways. The nomad, for instance, moves over familiar routes and returns to the same places, just as the line of the network must always connect with its end and starting points; mobility is routed, delimited and constrained. Other figures that underpin quite different ways of understanding mobility have not had the same restrictions. The meanings they are given connote not rigidity, permanence and determination but more of a fleeting wandering carrying the significance of a sort of freedom that moves along lines of flight without an end-point to stop it. In this section, I suggest that it is not the nodal point which dominates discussion, but it is rather the line – the connector – between these points which exemplifies nomadism.

Nomadism

The nomad has already figured in earlier geographical approaches towards mobility and the investigation of early modern societies. Their nomadism was tied up with a way of life that was incomprehensible and interpreted negatively. We saw how this could have very difficult implications for those on the end of sedentarist policy in Southern Africa (see the following chapter). On the other hand, nomadism has been imagined in a rather different light.

Linking mobility with power and fixity with geopolitical vulnerability, influential geographer Halford Mackinder argued how it was the geopolitical pivot of the Asian continent which offered the potential for far-reaching and powerful mobilities of force. In his historical study, Mackinder argued that 'all the settled margins of the Old World sooner or later felt the expansive force of mobile power originating in the steppe' (Mackinder 1904/1996: 545). The steppe was relatively isolated from sea power mobilities, while it offered a negotiable space of open desert plains. This permitted a brand of 'nomad power'. Mackinder traced the roving nomadic tribes of the region who had consistently overcome sedentary populations. The ability to move quickly and freely over this land could determine the conduct of war and the acquirement of power. He wrote,

> Is not the pivot region of the world's politics that vast area of Euro-Asia which is inaccessible to ships, but in antiquity lay open to the horse-riding nomads, and is to-day about to be covered with a network of railways? (1904/1996: 549).

In other words, the nomadic power of movement facilitated by the land afforded a set of structural conditions – an environment that would enable or determine a 'mobility of military and economic power of a far-reaching and yet limited character'.

Mackinder's dissatisfaction with the relative decline of British industry and empire saw him extol the virtues of thinking geographically. What he called geographical capacity was to be encouraged, for it gave the mind the ability to think cartographically in an expanded global sense. It was to flit easily across the map or to 'roam freely across the globe' (Morgan 2000: 59). Acquiring a mobile gaze (Parks 2005) could permit the greater appreciation of international relations and embrace the widest possible mobility of power (Mackinder 1904/1996: 546–547). Adopting the technologies of the railway and the aeroplane would guarantee Britain's success.

Mackinder clearly romanticizes the nomad's power for movement and the power that movement might supply. His approach marks mobility with meaning. Mobility meant power and power meant the promise of geopolitical dominance (Luke and O'Tuathail 2000). Here, movement is something to be feared and, in the right hands, it would be a powerful weapon. Mobility is the future, something to be attained in order to achieve. Curiously it is just these kinds of associations that more recent writers of the nomad have rallied around.

Philosophers Deleuze and Guattari's (1988) *Treatise on Nomadology* uses the nomad as a historic-contextual as well as a metaphorical notion to understand the mechanisms of two forms of power. They suggest that the power of the state is the power to subordinate fluids to paths and lines. To direct movement along 'pipes, embankments, which prevent turbulence, which constrain movement to and from one point to another'. In this way, movement is always 'dependent on the solid' and flows proceed, 'by parallel, laminar layers' (1988: 363). This form of power could be well visualized by the kinds of points, lines and networks discussed earlier on. The deployment of sedentary science was the deployment of this imaginary.

Key idea 2.3: NOMAD SCIENCE

Gilles Deleuze and Felix Guattari's *A Thousand Plateaus* (1988) is a remarkable book for the study of mobilities and the values that they attribute to it. In their well-used 'treatise on nomadology', the authors set about a sprawling chapter that explores two ways of thinking or *sciences* pitted against each other. Closely resembling the sedentarist and nomadic metaphysics detailed in this chapter, Deleuze and Guattari's discussion accounts for the labour relations found in the construction of Gothic cathedrals in the 12th century. Associations of journeymen or nomadic builders, from masons to carpenters, posed numerous problems for State management and regulation which they resisted through two forms of power: their mobility and their ability to strike. Taking the journeymen's mobility in particular, Deleuze and Guattari suggest that their struggle was composed of two diametrically opposing forms of power, the nomad artisan and that of the State architect.

In this sort of formulation, like Mackinder and Sauer, the nomad is not only reliant upon spatial features but is also set free by them. Deleuze and Guattari upturn the point-line network lens we have been discussing. Now it is movement that comes first, and then the line and then the point. And like Mackinder, this gives the nomad considerable emancipatory power. Elsewhere, Deleuze and Guattari's figuration has become incredibly popular as a symbol of freedom, plugging into other cultural stereotypes and contemporary ways of life. Recently Hardt and Negri's thesis on the system of *Empire* sees the forces that challenge that Empire to be unpredictable and especially mobile and fluidic – a mobile multitude where mobility and 'mass worker nomadism always express a refusal and search for liberation' (Hardt and Negri 2000: 212).

Further reading

(Deleuze and Guattari 1988; Kaplan 1996; Hardt and Negri 2000)

The nomadism of Deleuze and Guattari's workers turns the point-to-point networked formulations of the State (and perhaps spatial) science on their head. Although it seems that the nomad must move through points, they are not governed by them. Paths and lines of flight the nomads definitely take, but they do not do so slavishly. Points are merely fulfilled to get from one place to another. The points 'are strictly subordinated to the paths they determine, the reverse of what happens with the sedentary' (1988: 380). Every point is simply a 'relay' and 'exists only as a relay'. The path or the movement is, therefore, not dominated by the point, but it has an autonomy and a 'direction of its own'. The nomad goes from point to point only as a necessity and not as an aim.

This kind of nomad occupies quite a different space to those we have been presented with thus far. The nomad's space is not closed and regulated like the driver on the road or the highway. It is not 'striated' by walls and other enclosures. Rather, it is seen as an *open* or *smooth* space. The nomad's space is un-mediated by laws and controls for it is rather more direct; space and nomad subsist with one another. The nomad 'makes the desert no less than they are made by it'. The paths of the nomad are fixed in the sense that they are determined by a relation between nomad and the spaces in which they must return, such as the oasis or the water hole. But these move too. 'The sand desert has not only oases, which are like fixed points, but also rhizomatic vegetation that is temporary and shifts location according to local rains, bringing changes in the direction of the crossing' (1988: 382). Furthermore, the nomads' negotiation of this space works in a different way. Wayfinding is not done with any kind of global knowledge or distanced comprehension from nowhere, such as a representation like a map. Instead navigation is done by a localized and haptic engagement with somewhere. As 'sets of relations' from the 'winds, undulations of snow or sand, the song of the sand or the creaking of ice', these tactile qualities allow space to be crossed. The nomad space is 'localized and not delimited'.

Of course, the supposition that nomadism equals resistance springs up in more places than Deleuze and Guattari's treatment of artisans. The power of the nomad's mobility is often remarked upon as an important strategy in the evasion of power. Anthropologist Evans-Pritchard's (1949) study of the Sanussi–Bedouin resistance of the Italian Fascist occupation of Libya demonstrates the familiar romanticism of their mobility as a transgressive force. David Atkinson (1999) shows how the Bedouin enacted 'nomadic strategies' of fluidic movements that evaded the fixed point-to-point thinking of the Italian army. Quite simply nomadism provided an effective antidote to the orthodox strategies of the Italians. As Pritchard-Evans noted, the 'guerrilla imperative' was to 'strike suddenly, strike hard, get out quick' (Evans-Pritchard cited in Atkinson 1999).

Furthermore, the very flexible and mobile manner of 'nomad science' has not remained a weapon of the apparently powerless but is much more complex. Take architect Eyal Weizman (2002, 2003, 2007) who accounts for mobility in contemporary territorial conflict.

Case study 2.3: STRATEGIC POINTS AND FLEXIBLE LINES – THE GEOMETRY OF THE ISRAELI–PALESTINIAN CONFLICT

Eyal Weizman's pioneering work on the spatial architecture of the Israeli–Palestinian conflict demonstrates how the point and the line have emerged as key philosophical ideals from which to organize strategic mobility policy.

Tracking Arial Sharon's career from Chief of the Israel Defense Forces (IDF) Southern Command to Israeli Prime Minister, Weizman explores how Sharon proposed a defensive technique that enacted something quite different to the traditional strategy of linear points and lines that created a thin and fixed frontier. Sharon planned a series of relay points in order to compose a dynamic field of depth. Between these points 'mobile patrols' would be 'constantly and unpredictably on the move' (Weizman 2003: 174). Sharon's 1977 appointment as minister of agriculture that saw him undertake obligations for settlement policy allowed him to express this strategy further still. In order to construct a defensible territory that would 'consolidate' the Israeli Occupied Territories, the West Bank withstood the incursion of many points. Settlements composed the matrix of defence forming an open frontier that was not rigid but continuously fluid.

Between these points, numerous traffic arteries and road junctions were built forming IDF interchanges as observation and control points. Such points–line matrices operated as 'panoptic fortresses' orientated in their concentric circular rings to their outsides. In this light, these moving-creeping forms acted as weapons or wedges, slow-moving formations that occupied space.

Like the nomad, speed was of essence to Sharon's plans. As Weizman puts it, 'the axiom that the part to move faster across a battlefield is the one to win the battle' really counted here. A separation of speed occurred. The links connecting Israeli settlements and towns – the points – enabled fast velocities. Huge bypass roads allowed military and civilian vehicles to move quickly, while smaller dusty roads connecting Palestinian settlements made a similar journey take that much longer. The difference being 7.5 hrs to cross the West Bank.

Acting as 'synergistic systems', particularly within the most recent development of the West Bank security wall that divides Israeli from Palestinian land, the line and the point in this context interlock with one another. The point protects the line and its movement. The line – covered by various fortifications of wire, ditches, and dykes – protects the point's territory.

Further reading

(Graham 2003a; Weizman 2003; Coward 2006)

These forms of mobility could be contrasted with the figuration of the tramp or hobo popular within American cultural texts and music (Cresswell 2001). But at the same time, they can be brought into sharper contrast with the sorts of perspectives that occupied the previous section. More recent scholarship has romanticized these sorts of imaginings of the nomad as escapist, celebrated in a new kind of digital nomadism. Makimoto and Manners (1997), for instance, claim how digital technologies and the sorts of information communications systems we will look at in Chapter 5 emancipate their users from the soil of the earth. Mobile computing supposedly makes space unimportant, freeing up time and removing the need for face-to-face proximity (see Urry 2002, 2003). Discussions of the Internet have appeared enraptured by the ability to cast off the anchorage of one's body and identity. In the recently translated philosophy of Gaston Bachelard (1988), mobility is given the equivalence of imagination and is inextricably linked to freedom and liberty. '[M]obility is the liberating force,' writes Bachelard's translators, and that it is 'both the effect and cause of the exaltation of imagination which, by definition, is always active and dynamic' (ix). For Bachelard, thinking and imagining are *journeys*. Artistic creation enables this withdrawal and escape into the imagination, what he called an *invitation to journey*. A basis for other brands of nomadic thought which are far more political, in the context of cultural theory the nomad is deployed as an apt metaphor to enable one to move off-centre or, rather, to de-centre from established modes of thought and being (Braidotti 1994).

The flâneur, *the consumer and the tourist*

Another set of figures are usually cast in relation to the freedoms associated with the lines of nomadic flight. One in particular is usually an object of historicization, revealing the temporal contingency of economies, commodities, workforces and practices.

Representative of a mobile way of inhabiting the nineteenth- and early twentieth-century metropolitan city, the *flâneur* first really appears as an inhabitant of the urban realm within the physiologies of writers such as Balzac, Baudelaire and famously by Walter Benjamin (Benjamin 1973, 1985, 1986, 1999; Buck-Morss 1989). These were the experiences of authors trying to come to terms with what Berman calls the 'distinctive rhythms and timbres of nineteenth century modernity' (1983: 18). These rhythms beat to the drum of modernization, the industrialization of the city and the development of factories, railroads, and a market capable of 'everything except solidity and stability' (1983: 19). In Italy apprehensions of fluidity were further expressed in the futurists who took the technology of the aeroplane as their muse (Pascoe 2001). In Britain and America a swathe of writers captured the flux of a modernist imaginary, seen in the writings of Virginia Woolf, and the imagist poetry of Ford Madox Ford and Ezra Pound contemplating the London Underground (see Thacker 2003).

While Karl Marx (Berman 1982) sought to make people *feel* the abysses and violent eruptions of capital forces and social relations, the *flâneur* told of a life on the ground – what it meant to inhabit these societies, how one dwelt on the Parisian streets and interior spaces of the arcades. To *flâneur* was to move – quite lazily so – against the flow of contemporary existence, to enact a kind of detachment in order to ruminate upon encounters and experiences.

Detachment meant brief and fleeting meetings. Nothing was solid for the pedestrian who moved from relationship to relationship. Writing on Baudelaire, Benjamin notes how his behaviour was almost entirely asocial. Detachment meant disengagement and downright critique. The *flâneur* was characterized by *his* mobility which seemed lackadaisical and relaxed in opposition to the frenetic activity of modern life. Benjamin's *flâneur* became associated with a manner of silent protest that rejected the mass industrialization of the city and the new working classes. He was simply, 'unwilling to forego the life of a gentleman of leisure'.

Key idea 2.4: BOTANIZING ON THE ASPHALT, WALTER BENJAMIN, *FLÂNERIE* AND THE ARCADES

Marxist philosopher Walter Benjamin's writings have become some of the most famous and influential words on the *flâneur*. Treading the streets and researching at the Librairie Nationale during his stays in war-torn Paris, Benjamin saw the emergence of *flânerie* as an expression of discontent and a rejection of modern existence. He gathered together these writings in several volumes, the unfinished *Arcades Project* (1999) being an amazing collection of notes, observations, thoughts and quotes.

Paris itself offered up a landscape that was especially conducive to mobile practices. As Benjamin states, 'Strolling could hardly have assumed the importance it did without the arcades.' A product and symbol of technological invention in glass and iron work, the arcades presented both a home and laboratory for the *flâneur* occupation. The arcades offered little worlds, a 'city, even a world, in miniature' where the *flâneur* could stroll with other like-minded people and make comparisons with the world outside its doors.

Like the arcades the streets were the home of the *flâneur*, too. They housed the spectacle of movement and the spectacle of the commodity for the *flâneur* to indulge his passion of thought and contemplation.

> The walls are the desk against which he presses his notebooks, newsstands are his libraries and the terraces of cafes are the balconies from which he looks down on his household after his work is done.
>
> (Benjamin 1973: 37)

But it was mobility that most characterized Benjamin's figure. Playing the detective or observer of behaviour, strolling permits the *flâneur* with the 'the best prospects of doing so'. Strolling gave one anonymity while it disguised his practice of study. Moving with the crowd developed the *flâneur*'s powers of awareness, instinct and observation: 'He catches things in flight' (Benjamin 1973: 41).

Further reading

(Frisby 1985; Tester 1994; Benjamin 1999)

Resembling the evasive power of the nomad, Benjamin tracks how *flânerie* became as much an art of escape from power as it was a way of detaching oneself from modern life. Bemoaning the increasing regulative rules of the Parisian population, for women as well as men, Benjamin saw how Baudelaire's continual mobility helped him to fly from his creditors and avoid his landlord, 'the city . . . had long since ceased to be home for the *flâneur*' (1973: 47). The destruction of the *flâneur* was made complete with Haussmann's invention of the Parisian boulevards and the emergence of street traffic. Traffic demanded a new and modern experience of the city. Entirely new ways of dwelling and negotiating the moving chaos saw the development of mental and bodily capacities. Reading Baudelaire's descriptions, Berman shows how the Parisian 'suddenly thrown into this maelstrom', must not only learn new but also rely upon old bodily movements simply in order to stay alive. The capitalist ethos of 'survival of the fittest' finds its expression in the act of crossing of the street – in the formation of 'sudden, abrupt, jagged twists and shifts'. Learning to live with traffic meant attuning to 'its moves', to 'learn to not merely keep up with it but to stay at least a step ahead' (Berman 1982: 159).

For others, the *flâneur* was more than a simple observer, silently protesting while carelessly detached from what was going on. The *flâneur* 'is also purposive seeking to unravel the mysteries of social relations and of the city, seeking to penetrate the fetish'. According to Harvey, Balzac demonstrated a 'perpetual urging to check the city out and find things out for oneself . . . "Look around you" as you "make your way through that huge stucco cage, that human beehive with black tunnels marking its sections, and follow the ramifications of the idea which moves, stirs and ferments inside it"' (Balzac in Harvey 2003: 57).

Mobile method 2.3: *FLÂNERIE* AS MOBILE ETHNOGRAPHY

Even as the *flâneur* tells us many things about the mobile inhabitations of cities and attempts to study them, Chris Jenks and Tiago Neves have suggested the *flâneur* provides a framework through which to critically reflect on the methodologies of urban ethnographic practice.

Defining the *flâneur* as 'one who walks without haste, at random, abandoning himself to the impressions and sights of the moment' (Jenks and Neves 2000: 1), the '*flâneur* introduces a phenomenology of the urban built around the issues of the fragmentation of experience and commodification' (2000: 1–2).

The wandering observations actually provide a very useful model for conducting our own kinds of observations on the rhythms and movements of places like cities. Experiencing the fluxes and flows of the everyday – as *flâneur* – may mean watching people's movements, gestures and comportment. It could suggest paying considerable attention to how people use and consume public spaces.

It may mean taking part in the pacing of the city itself. 'Empathy is the nature of the intoxication to which the *flâneur* abandons himself in the crowd' wrote Benjamin (1973: 55). In this way, becoming a *flâneur* may mean stepping into the shoes of those he/she walks with. For Marx, one could fulfil the role of a ghost lusting for corporeal possession, 'Like a roving soul in search of a body, he enters another person whenever he wishes' (Marx in Benjamin 1973: 55). Flitting in and out of these positions could mean performing the role of the subjects one is researching. By taking part in their activities and practices, one may build a personal perspective on the visual and sensorial consumption of the city, for example.

In practice

- Pay attention to micro bodily movements and practices of inhabitation and consumption
- Question how you may perform the mobile roles of your research subjects

Further reading

(Benjamin 1973; Jenks and Neves 2000; Pink 2008)

Like the *flâneur*, the consumer has been constructed as a similarly mobile stroller. As various writers (Bowlby 2001) note in their studies of consumption in modernity, 'Shopping is playing at mobility,' involving a playfulness, a sense of wanderlust, 'it is the parody of mobility as perpetual, happy, directionless to and fro' (2001: 25). The figure of the consumer has become the symbol of the ultimate global nomad or citizen. It is the consumers 'who really have the power' writes economist and business guru Kenichi Ohmae (1990: xi) in his treatise on the borderless economy. Given the command of new information and data flows, the consumer is no longer duped into immobility. The mobile consumer, who votes with his or her very mobile feet, appears to drag the multinational corporation along in its wake. Business is merely the servant to the interests of the customer. Ohmae even goes as far as to suggest that the mobile consumer *should* become the model for international business for whom borders matter very little. Not surprisingly then, shopping and consumption are analogous to mobility and freedom (Gudis 2004; Cronin 2006, 2008). For some, this is a postmodern form of *flânerie* where one drifts along seduced by the array of consumer goods, advertisements and symbols supplying new wants that were not there before (Chambers 1986).

As we saw earlier in historical context, the rise of Western consumerism arose with increasing freedoms for women to depart from the public sphere into environments made almost exclusively for them (Friedberg 1993; Domosh 1996, 2001). Moreover, the actual practice of shopping enacts what Bowlby (2001) has called the 'antithesis of property', or immobility. 'In this sense', she writes, 'it represents a pure mobility of selves and objects.' Shopping and buying involves what she describes as a series of instants moving on. In buying one may purchase an item and a change of lifestyle. Shopping is thus also a movement between different identities – the sort that one has at their disposal, and the one bought through the transaction. Buying, therefore, involves a social mobility as well as a metaphorical or geographical one. Furthermore, it is mobility that characterizes the shopper for those who hope to lure, tempt or convince the consumer to stop and buy their products. The consumer is 'in a continual state of passing through or passing by walking up and down the supermarket aisles, browsing through the newspaper, driving through town' (2001: 217).

The life of the tourist, has unsurprisingly come to stand for a certain sort of modern existence – compelled to search out new and authentic experiences (MacCannell 1992). John Urry's comparison of the tourist with the pilgrim highlights the 'worship' of signs and places which are sacred and come to compose the gaining of an 'uplifting experience' (Urry 1990: 10). Everyday Western mobilities bear more than a passing resemblance. The tourist's search for authenticity may always be faced with failure; any notion of finding the pure rootedness of a place will almost never be met. Paul Fussell's (1980) analysis of travel in literature explains how tourists find themselves continuously in *pseudo places* on the way to places that are in themselves sites whose function is to primarily sell things to tourists. Airports act as stepping stones in between, permitting 'instant recognition' and orientation (see Chapter 5). Furthermore, the tourist may also want to *escape* the 'inauthentic' themed simulacrum of the shopping mall or the theme park. The tourist and the holiday they buy may well signify a personal letting go of everyday life in order to consume the place, and the life it might bring, of the countryside, nature or elsewhere (Crawford 1994).

The emancipatory significance of travelling appears to be no less important for interpretations of filmic pleasure. The moving cinema camera becomes 'a means of transport' (Bruno 2002: 24). For Wolff, the cinema has afforded women that ability to *flâneur*, to loiter, to stroll to wander without fear. 'The "spectatrix" could thus enter the world of a *flâneur* and derive its pleasure through filmic motions' (Bruno cited in Wolff 2006: 21). In addition to the movement itself, film provides access to new horizons of destinations, and as Bruno writes, 'the female spectator,' 'a *flâneuse* – travelled along sites' (2002: 17). Benjamin (1985) writes how film unlocked people who could 'now, in the midst of its far-flung ruins and debris', 'calmly and adventurously go travelling'.

The apparently liberatory touring and 'mobile gaze' of the film spectator is even transferred into other domains such as architectural design. Blending the value of the nomad, the *flâneur* and the site-seeing tourist, cinematic visualizations both inspire and become materialized in architecture. Inspired by film director Sergio Eisenstein's notion of montage, various architects have sought to create cinematic-like experiences in their buildings. Rem Koolhass's idea for the Rotterdam Kunsthal was conceived as a sort of serial vision, the route structured like a plot for a film. The visitor teleologically moves their way through the museum being bombarded by different views and visual stimuli at carefully chosen points in time, giving a definite and linear filmic narrative with a 'beginning, middle and a powerful climax' (Porter 1997: 115). By this method, architectural design is not only conceived as a visual passage but is also formulated through one; fly-throughs and virtual pre-visualizations release the gaze from the limitations of its body.

Fluidity and fixity: a synthesis

What should be evident by now is that even though sedentarism and nomadism appear to be at odds with each other, they seldom work as simplistically as that. Similar metaphors and analogies show up in both points of view and shift in emphasis depending on the context in which they are used. In one situation, nomadic movements were treated with disdain and fear, and in another, the nomad became a hero of democracy and freedom.

Take the figure of the *flâneur* once more. For others, the drifting tendencies of their activity are actually far more anchored and grounded (Buck-Morss 1989; Wolff 1993). While the *flâneur* is laden with the qualities of expressive freedom and thoughtful detachment, for women *flanêrie* meant a practice they were to be excluded from. *Flanêrie* meant mobility for some and immobility for others, relegating women to the sphere of the home. Pollock writes how even for bourgeois women, 'going into town mingling with crowds . . . was morally dangerous'. Mobility in the city became an issue of morality and reputation. Public space was where 'one risked losing one's virtue' and 'going out in public and the idea of disgrace were closely allied' (Pollock in D'Souza and McDonough 2006: 7). Women's practices were finely striated according to divisions of race, class and gender, inscribed with certain ways of acting and being seen out (Domosh 2001). Even more recent studies show how consumption is far more tethered to enduring forms of relationship than it is marked by superficial and fleeting encounters. Examples including analyses of the contemporary shopping mall (Shields 1990; Miller *et al.* 1998; Miller 2001b), and Nead's (2000) investigations of the department store in Victorian London, show how mobile consumer practices are very much grounded within the obligations and expectations of family ties and kinship as well as other social relationships.

This re-clarification of the *flâneur* chimes with Peter Geschiere and Birgit Meyer (1998) who outline how the two intellectual projects I have discussed seem devoted to either a process of grasping and fixing – holding down to immobilities and fixities, or one of celebrating fluidity and dynamics. In this section of the chapter, we can now turn to other approaches that don't really fit either sedentary or nomadic viewpoints. In this section, we explore how the underpinning logics of both fixity and fluidity are paradoxically interrelated.

While there are many different examples of this, we will now take a look at research which has attempted to move beyond sedentarist and nomadic metaphysical viewpoints in order to understand the fluidic and fixed characteristics of places as they relate to one another.

Mobile method 2.4: MOBILE ETHNOGRAPHY AND MULTI-SITED RESEARCH

From anthropology, ethnographic research methods provide one avenue that seeks to uncover lives as they are lived. Ethnographic research seeks to explore the richly detailed and complex life-worlds of its respondents. This process, however, has traditionally been viewed as an immobile and rooted approach, stuck within the confines of specific cultural contexts or single locales. Moves afoot within the anthropology of the late 1980s sought to find ways to attend to the circulation of 'cultural meanings, objects and identities in diffuse space–time' which could not be accounted for by 'remaining focused on a single site of intensive investigation' (Marcus 1998: 80).

In a series of influential writings George Marcus drove the turn for multi-sited or transnational research and field methods which strove to track and trace 'unexpected trajectories' both 'within and across multiple sites of activity'. Marcus argued for ethnographic research methodologies that could account for different places and the *interconnections* between them.

This meant the familiar deep understanding of a place while exploring the relationships, associations and translations between those sites of enquiry tied together by 'chains, paths and threads' through which the ethnographer travels (105). Speaking to the kaleidoscope of research we will explore on migration, objects, metaphors, even stories, lives and conflicts, *following* mobiles through multiple sites and spaces emerges as the key practice of

Marcus' suggestions. *Follow the people* is an obvious strategy wherein the movements of migrant communities and diasporas are followed and moved-with. *Follow the thing* has been similarly articulated through a body of research on material cultures and practices of science (Latour and Woolgar 1979), and even the consumption of food discussed previously (Cook 2004).

In practice

- An in-depth ethnography of a place may not be sufficient for a research object and/or subject on the move
- Studies of migration particularly may require multi-sited research in order to account sufficiently for their subject's dwelling in new social and material settings
- Attention to the interconnections between places (such as communications, letters, emails, gifts) can highlight the maintenance of links and social groupings (see later in the chapter on translocal places)

Further reading

(Marcus 1995; Marcus 1998; Cook 2004)

Stabilities

Returning to Tuan we might reconsider whether our immediate tarring of his work with the brush of sedentarism was a little hasty (1974, 1977). Tuan suggested that nomadic groups do, in fact, create their own attachments to places in two main ways. First, he purports that nomads may find attachment to the places that they continually return to. Nomads temporarily find pauses of place and home before leaving to their winter camp and again returning to their summer location. The points on Tuan's metaphysical map, therefore, gain familiarity and are subject to the welcome of an arrival and also the tear of a departure. Tuan explains how it is not necessarily absolute spatial coordinates that matter in the construction of meaning, but the stabilization of encounters (1978).

Second, Tuan notes how nomads are able to turn the line or the space between places into place. Repetition is important because nomadic movement tends to be cyclical. The line or the vector between places

(1978: 14) 'will be followed year after year with little change', he says. Through this cyclical repetition, 'the path itself and the territory it circumscribes are likely to acquire the feel of place. Space, as it gains familiarity and meaning, is barely distinguishable from place' (1978: 14). From this point of view, mobilities are not the simple aggressor to any chance of meaningful attachment to places. Rather it is the immobility of their repetition that habituates into a stable investment of meaning. The line of the route followed can be as meaningful and significant as the end-points of the journey.

J. B. Jackson fails to stick with this interpretation either and develops an approach more sensitive to both nomadism and sedentarism. He concludes that the 'the real significance of the temporary dwelling, of the box house . . . I think it has always offered . . . a kind of freedom we often undervalue' (1986: 100). The trailer park lifestyle allowed an emancipation from ties; emotional ties to land, to people, to property and things, to a community. Mobility meant escape.

So we could suggest that Tuan, and also Jackson, are not nearly as adverse to the idea of mobility as first thought. Points and places are important, but so is the movement between them which can become just as significant and as familiar as the point. There is still a sense of immobility there. Mobilities are occurring, but there is no diversion from the line. What is important is the stability of these mobilities – the repetition which formulates attachment. Tuan muses on the life of a high-income executive whose

> relation between mobility and place can be very complicated. . . . The home in the suburb is a place, it is perhaps also a showplace in which lavish entertainments occur. It is a workplace, for the busy executive brings his work home. . . . The office is a workplace but it is also the executive's home – to the extent that it is the centre of his life; he may have an apartment in the office building or downtown where he occasionally spends the night. . . . To compound the complexity the circuits of movement and their resting places in any period of years represent only a stage in the executive's upward mobile career. At any stage a routine of movement between places is established; the stage itself, however, may be viewed as a "place" in the sense of a pause in the executive's rise to the summit of his profession.
>
> (1978: 15)

Mobility does not necessarily serve to threaten an attachment to place. For a route well travelled may over time turn into a meaningful place, just like the places or the nodes at either end of the route. Repetition is key, 'Repetition is of the essence,' 'repeated experience: the feel of place gets under our skin in the course of day-to-day contact.' We will come to this in later chapters, but Tuan suggests that 'The feel of the pavement, the smell of the evening air, and the colour of autumn foliage become, through long acquaintance, extensions of ourselves – not just a stage but supporting actors in the human drama' (1974: 242). It is the 'functional pattern' of people's lives that (1974: 242) is crucial to the formation of senses of place. Thus by carrying out daily routines, habits and regular movements, one may follow established paths, the result being that 'a web of nodes and their links is imprinted in our perceptual systems and affects our bodily expectations' (1974: 242).

While Tuan found stability in rhythms and routines of movement, he also found that for those on the move, anchorage and permanence could be instilled within things and objects they travelled *with*. He writes how English gypsies are strong collectors of various kinds of objects from china to old family photographs. This 'emotional anchoring' enables a mooring of meaning and subjective feeling towards 'wherever they happen to be'. In fact, it is the process of travelling with others that enables feelings of place while on the move. Comparing gypsies and young lovers, Tuan suggests that they are 'placeless in both senses of the word and they do not much care' (1974: 242). Ethnographic investigations of gypsy traveller communities demonstrate similar findings to Tuan. For a respondent of Kevin Hetherington's 'Everybody wants to be part of something. . . . People want to know that they are part of something larger like that they also want to know where it is that they fit in' (Hetherington 2000b: 83).

From space to place

Anthropologists Geschiere and Meyer (1998) warn against the tendency to follow one path wholeheartedly and suggest that even within the most fluid and ephemeral processes of globalization there may be found 'constant efforts towards closure and fixing at all levels'. The authors argue that thinking about fixing encourages us to ask new questions about 'who creates new boundaries and securities', why these fixings might be created and with what effect and 'against or with whom' (Gescniere and Meyer 1998: 614). In other words, practices of fixing and closure are employed by people as a way to make sense of – a way to live – within globalization. Consider the authors' words of caution, that this 'intellectual project of fixing' threatens to 'overstress' an idea of stability with the consequences that 'identity only appears as an antipode to global dynamics, as a means to create closure' (1998: 614).

David Harvey's work has been drawn on regularly within the literature to exemplify the conceptualization of fixity in relation to mobility (Cresswell 2004). In his *Justice, Nature and the Geography of Difference* (1996), Harvey examines the place-making strategies surrounding the town of Guilford, Baltimore, in the United States. Harvey notes the various streams and flows of information, capital, people and raw materials that have criss-crossed the town. Paradoxically, these flows have been interpreted in vastly different ways by the inhabitants of Guilford. In the event that two elderly citizens were found murdered, the African American, lower-class, and migrant population of the town had the finger pointed solely at them.

As a town suffering from the decline of its manufacturing industries Harvey examines how local authorities and industries sought to attract incoming flows of investment capital in order to stimulate growth and regeneration. These sorts of mobilities were not threatening but essential to the survival and prosperity of the town. The two reactions saw very different sorts of significance applied to mobility, yet both event and situation were apprehended with strategies of boundary and place making. Attempts to make Guilford meaningful as a distinctive place, a point, were made. Harvey asks, 'So what kind of *place* is Guilford? It has a name, a boundary, and distinctive social and physical qualities. It has achieved a certain kind of "permanence" in the midst of the fluxes and flows of urban life' (Harvey 1996: 293). Those concerned with the murder of the elderly couple, reinforcing the idea of Guilford as a bounded and meaningful locality with historical and local character, challenged the movements and right to belong of African Americans or suspicious migrant workers. Those concerned with the economic future of the town attempted to do the same by manufacturing a Guilford with distinctive qualities – to create a recognizable *place* – a place with that might compete more successfully for mobile capital.

Harvey has targeted what can best be described as a reactionary response to the effect of time–space compression – strategies to trap or mark out a place within the uncompromising fluidities of a de-industrializing economy. Mobility is held down; a box is drawn round place in order to make it secure and viable as an economic competitor. Only once Guilford was stabilized as a relatively bounded and meaningful entity could life begin again. Harvey is envisaging neither a world of simple fixities nor one where flows are stopped for eternity. These are not simple immobilities but rather relative permanencies, an explosion of 'opposed sentiments and tendencies' that have been caught and crystallized into a fragile truce between fixity and motion. Mobilities, as he suggests, are vital to the continual renewal or the permanencies created. The boundary of Guilford was a semi-permeable membrane. It was bold in its solidity and distinctiveness in order to hold off suspicious migrants and to sustain the town's economic prospect, while it was open enough to encourage the filtration of capital once it had been secured.

Progressive places

Harvey's view is not universal and has been subjected to serious critique from Doreen Massey (Massey 1993). Arguing that Harvey relies upon a rather 'reactionary' conception of place, Massey shows how Harvey seems to equate the mobilities of time–space compression with fluidic metaphors that conjure up images of a boiling soup of uncertainty, ephemerality and volatility that surrounds and bombards the world's inhabitants. Massey encourages us to look past the sense that 'real' meanings of places can only be found in fixity and rootedness, or indeed, that they are simply a reaction to the hubbub and fluidities of globalization.

Resembling Simmel's writings of the experience of the city, Harvey supposes the construction of at least temporarily protecting circles, circles that form the individual's 'milieu', circles or boundaries that protect against the outside, guarding 'the achievements, the conduct of life, and the outlook of the individual' (Simmel and Wolff 1950: 417).

Alternatively, Massey (1993) wants to put forward an idea of meaningful places, and meaningful activities in a way that is bound up in permanent mobility. In what has become a famous muse on walking along her own local high street, Massey asks us to consider the very mobile activities that constitute meaningful places: 'Thread your way through the often almost stationary traffic diagonally across the road,' she writes, 'and there's a shop which as long as I can remember has displayed saris in the window.' Massey goes on to discuss the migrant lives going on under a regular flight path to London Heathrow (Massey 1993: 153).

While Harvey renders meaning making as efforts to temporarily stop and bound off places for various personal and political projects, Massey does something a little different. Both authors seem to presuppose a given flux to the world, of people, goods and especially capital. But Massey suggests that meaningful places are always moving on. They are not temporarily stopped but exhibit a sense of momentariness, what she later describes as 'thrown togetherness' (2005). Conceiving places in more extroverted manner, Massey (1993) envisages 'networks of social relations and understandings' that occur beyond the now and the moment of a place. These 'relations', 'experiences' and 'understandings' are actually constructed at a level and scale beyond the place of the home or the street. From this sort of perspective, though the idea of the point-as-significant-place is more than a temporary fixity, it is a non-entity because the point never exists. It is conceived in thought and through efforts to close down place (which we will discuss later). Places and meaningful activities are really constituted through ephemeral practices along much wider and extensive networks of flows and mobilities.

It seems to be key that place involves the challenge 'of negotiating a here and now (itself drawing on a history and a geography of thens and theres); and a negotiation which must *take place*' (Massey 2005: 140, emphasis added). What 'must *take place*': this is so important as we begin to see place as more than an immobile fixing of what has gone on, but a verb or a doing (Merriman 2004). Consider how even the most fluidic and temporary of places, such as airports and particularly the motel (see the box below), characterize an ambivalence between mobility and immobility (Bechmann 2004; Normark 2006).

Case study 2.4: INTENSITIES OF STAYING AT HENRY PARKES MOTEL

Meaghan Morris's (1988) classic article on Henry Parkes Motel supplies us with a conception of place which at first appears schizophrenic in its attitude to mobility. With its signs of tourism and moteldom, foyer and rooms, the motel is at once a symbol of coming and going, while it is a family home evident only a few yards away from the front office.

At the same time as being a motel and a home, the establishment acts as a kind of local community centre serving the local leisure economy with its gym and sports centre and swimming pool. For Morris this creates even more confusion in defining the motel. Its space is continually intermingled by different people, blending domestic, local and personal uses with more 'passing' people and their needs. According to Morris (1988: 7), 'The motel's solidity as *place* is founded by its flexibility as *frame* for varying practices of space, time – and speed.'

But where Morris pushes on the discussion in the most interesting way is perhaps in her denial of the kinds of dichotomies and dualisms usually set up in discussions of place. The Henry Parkes achieves a coherence of place even though it is in continual motion because of its 'durable familialism'. The mobilities of the Henry Parkes are multiple; they are touristic, neighbourly and proprietal. And while these might have been positioned as direct opposites by others, for Morris they are merely divided by 'degrees of duration', or what she calls, "intensities of 'staying' (temporary/intermittent/permanent)" (Morris 1988: 8).

Further reading

(Morris 1988; Normark 2006)

Translocal places

The end-point I want to take of this reworking of meaningful mobilities can be found in work concerned with migration and the process of what anthropologist Arjun Appadurai (1995) terms trans-locality. Redressing our conceptions of place with more mobile understandings of their extensiveness, transitoriness (or persistence) and their permeability, we can look at how migrant diasporic groups forge places between and away from their places of departure.

Mobile method 2.5: EMPLACING MIGRANT SUBJECTS

While we have witnessed a methodological turn to multi-sited ethnographic research methods sympathetic to the mobile existence of migrants, concerns have shown that this kind of approach could render the 'site' of research enquiry as a hyper-mobile 'space of flows' and, therefore, evacuated of historically mediating contextual forces.

Migration scholar Michael Peter Smith urges researchers to engage methodologically more fully in the *emplacement* of mobile subjects.

For Peter Smith, mobile research should continue to focus upon border crossings and mobilities of its subjects, while examining the importance of the specific contexts in which these actors move through and act. As he writes, 'the actors are still classed, raced, and gendered bodies in motion in specific historical contexts, within certain political formations and spaces' (Smith 2005: 238).

Of clear importance to researchers more attuned to questions of space and place, migration scholarship examines the importance of location, place and the geography of migrant subjects – from their homes to sites of worship (Blunt 2005; Blunt and Dowling 2006). Anne-Marie Fortier's (2000) research on *Migrant Belongings*, for instance, explicitly discusses the communal church settings in which much of her research took place. In addition to mediating the research encounter, Fortier notes how such settings play different roles in the 'community life', and thus to render them anonymous would be to lose 'all their vitality'. Similarly, Latham and Conradson advise turning to the 'panoply' of spatially situated 'mundane efforts' that enable transnational mobilities within particular places (Conradson and Latham 2005: 228).

In practice

- It is important to consider the points and places in which and through which mobilities pass
- To examine how these contexts provide social and political contexts may show how particular mobile bodies as classed, gendered or raced are constructed

Further reading

(Fortier 2000; Smith 2001; Conradson and Latham 2005, 2007)

Figure 2.5 Filipino translocal subjects

Source: Copyright © McKay (2006)

Examined in more detail in the case study box below (illustrated above), geographer and anthropologist Deirdre McKay (2006; Conradson and McKay 2007) maintains that the formation of migrant subjectivities has been recognized as either one of 'fixing' and 'placing', whereby one becomes 'part of' the places in which they are located. Alternatively, they may enact a process of deterritorialization, wherein migrant lives are placed in continual reference to their diaspora and to their place of origin. For McKay (2006), such an understanding begets the possibility for more mobile subjectivities that do not 'necessarily lock one into a singular locality'. Understood in a more plural fashion means not a choice between 'placing' and 'deteritorialization' but the potential for both. Reterritorialization exudes the possibility of *translocal* places that extend the sociality of home into somewhere else. These are places that may travel or circulate; they are places brought with one and recreated in new contexts. And they are places that occur along and across, formed through routes of movement created between the site of departure and the new locality.

In this sense, diasporas form place-like attachments along extended social networks, or what anthropologist Pnina Werbner (1990, 1999) describes as pathways. These spatial extensions may see diasporic networks knitted together through practices of gift giving and exchange in order to form material and mobile cultures of national identity (Tolia-Kelly 2008). Following Marilyn Strathern's (Strathern 1991: 117) description of how migrants 'make the places travel', Werbner tracks how Pakistani migrants in Britain import commercial goods producing a traffic of 'sentimentally loaded, ceremonial exchange objects' such as food, cosmetics and jewellery (1999: 25) and clothing. Common among other South Asian diasporic communities it is the traffic in particularly clothing and fashion goods (Jackson *et al.* 2007), namely, 'objects–persons–places–sentiments which is one of the most significant bridges of distance spanning global diasporic communities and transnational families' (Werbner 1999: 26). For other post-colonial migrants of South Asia and East Africa, the presence and display of photos, pictures and paintings can presence their migratory experiences (Tolia-Kelly 2004, 2006).

Case study 2.5: PUTTING PLACE ON THE LINE, FILIPINO TRANSLOCALITIES IN HONG KONG

McKay's ethnographic study of Filipino migrant workers in Hong Kong provides a fascinating case study of Appadurai's notion of translocalism. As McKay argues, part of this process is the transportation and reconstruction of places elsewhere. '[M]obility can,' McKay suggests, 'in fact, recreate elsewhere as a part of "home" through the reterritorialisation of locality in its extension into a global world'.

A prime example of this reterritorialization is the event space of the migrant workers that converge on Statue Square and the Plaza area under the HSBC bank building after church services on a Sunday morning and afternoon. The urban plaza and square are transmuted as a site of migrant interaction resembling social gatherings enacted at home. Filipinas swap 'news, gossip, food and money', on blankets set out at prearranged locations. The Hong Kong centre has temporarily become a re-creation of the villages the Filipino migrants have left. In this light the place of home works in multiple, it exists in separate locations, albeit slightly differently and temporarily. At the same time, the place of the Filipino village becomes, paradoxically, something that is enacted at the same time. For McKay, translocal place is simultaneous and stretched out. Associations and attachments are maintained during the Sunday social gatherings. Migrants regularly use this time to text message and call home in order to catch up on weekly news from their friends and families, to monitor their children and child-carers, to check up on investments and businesses they may have left.

Further reading

(Appadurai 1995; McKay 2006)

CONCLUSION

In this chapter we have seen how meaning is part and parcel of mobility. Mobilities are inscribed with significance and ideas. Perhaps mobility is always meaningful and, therefore, never simply movement. Even approaches that characterize mobility in the most abstract of ways see mobilities imbued with value-laden judgements and labels. Importantly, these values, judgements and meanings really do matter. They matter for the way mobilities have been understood and treated within a discipline like geography, and in just the same way, they matter for the interpretation and treatment of mobilities in society.

As we saw, this societal and academic context is intertwined in a way which reflects the importance of context in the making of meaningful mobilities. The visions of mobility painted by early geographers reflected dominant societal attitudes. Corbusier's inspiration showed how figures, metaphors and even the simple shape of the movement of a river may be taken to be much more than an innocent representation. The approaches of the 1970s spatial scientists were attuned to Cold-War needs and attitudes. These were clearly useful representations of mobility that enabled quite complex associations and flows of movement to be visualized, providing highly enduring frames through which mobilities have been often understood. The academic address of mobility, therefore, tells us a great deal about how specific societal contexts shape the way mobilities are treated and understood in the wider world. Thus, in some instances, mobility may be registered as a positive sign of progress and wealth such as the shopper or the consumer, while in another it is addressed as dirty and backwards. The meaningful and ideological coding of mobility can reflect contextual societal attitudes and social practices.

From military strategy to the vilification of Southern African migrants, these sorts of meanings, discourses and models may, completely unquestioned, reproduce themselves by acting as norms, or ideals as 'the metaphor with which we work' is taken very much, 'for granted' (Barnes 2008: 134). The planners of spatial science found that the models they developed could be seen in the world. Paths, lines or channels were apparent as physical structures like highways or railway tracks, or they could be seen in organizational arrangements such as the regional structure of settlements.

In the following chapter, we will examine what happens when these ideological codings become norms against which mobile bodies are formally judged. The consequences of imagining mobilities in the way we saw spatial scientists do, as productive and rational billiard-ball-like atoms that behave according to simple economic and physical laws, (Imrie 2000) are investigated in terms of the problematic real-world effects upon those who do not conform to this imagining. As we sketch out the relationship between mobility and politics, we consider how several of the dominant figures of mobility often meet each other head-on with exclusionary and sometimes violent implications.

3

POLITICS

Any human movement, whether it springs from an intellectual or even a natural impulse, is impeded in its unfolding.

(Benjamin 1985: 59)

The only form of resistance is to move.

(Harvey 2005: 42)

INTRODUCTION

Consider the relatively individualistic and autonomous act of driving down a road. Before we can even get going, just as one feels the gentle click of the seatbelt into its socket, one's sense of freedom and expression may immediately slip away. The driver must consciously or unconsciously be directed by a host of limits and directions that he or she must bear in mind (although these depend greatly on the context: see Tim Edensor's account of Indian motoring in Chapter 5). The limits of one's speed might be set by a general 'speed limit' which is reminded to us by the technology of speed limit signs, cameras and road bumps. Peter Merriman's (2005a, 2006a, 2006b, 2007) rich accounts of driver regulation in the context of post-war Britain highlight a host of rules and regulations that discipline the driver. In the United Kingdom it is illegal to drive without a seatbelt fastened correctly.

From national 'rules of the road', to more localized restrictions and nuances, the driver's journey is not simple at all. Take driving from a particular moment in time. In 2001 a fuel shortage in the United Kingdom saw petrol stations deluged by anxious car drivers seeking to fill up their cars. At the same time mass protests were undertaken by angry lorry drivers who conducted go-slow convoys in order to clog up dual carriageways and motorways. In this instance the idea of the autonomous driver is questioned again. The road provided a visible public space and a forum for contestation and opposition to the way things were being handled. The acts of drivers impeded on others who were driving with them. Other drivers felt the effect of the queues some miles further up the motorway. And viewers of news broadcasts and newspapers caught up with the events simultaneously or later in time.

Finally, take another moment in the example of the death of the pedestrian, Cynthia Wiggens, as she tried to cross a busy highway (see Graham and Marvin 2001 for more detail). Wiggens was an employee of the Walden Galleria Mall, situated on the edge of Buffalo in the United States in 1995. Cynthia Wiggens worked at the mall and required the use of the public transport system in order to get there. The rub was, the stop of the bus service she had to use was situated across a seven-lane highway from the mall. While the mall had been designed to accommodate private automobiles, bus services had not been at the forefront of the developer's plans. On 14 December 1995, Wiggens was run down by a 10-ton truck while she tried to make her twice daily crossing from the bus-stop to the mall (Graham and Marvin 2001: vi).

What do all of these three issues have in common? The answer that this chapter will deal with is politics.

Political engagement, discussion, participation and decision making, set and define the roads we drive on and the laws we must abide by. At the same time, politics means that the possibility of particular roads and the laws we must abide by, or the kind of car we might choose to drive, may be contested and opposed. It is politics that makes car travel possible while it also regulates it, shapes it and may work to halt or change it. Moreover, mobility provides a space for a politics and renders our ability to be *political* by shaping one's capacity to contest, deliberate and oppose. Think for instance of your own access to mobility. Can you drive? Do you drive? Can you afford to drive? Do you use public transport? Do you take walking down the street for granted or find it a struggle? How are you placed in relation to mobility? These questions are intended to provoke the thought that different people are placed in very different ways to mobility. And across all of these aspects, our ideas about mobility, related to the sorts of meanings discussed in the previous chapter, colour the way mobilities are interpreted, understood and treated.

Already we have three or four main components to our understanding of mobility and its relation to politics. While clearly all sorts of mobility intersect politics in one way or another, this chapter will trace out how these relations have been understood broadly through a host of different examples and spaces. The chapter is structured as follows:

First, the chapter will set out these entangled components of mobility in the following section. The chapter attempts to untangle what are very knotty issues of mobilities and politics, before exploring how these themes play out in some of the dominant areas of citizenship, disability politics and others. Finally, the chapter ends with a consideration of how mobilities offer up a political space wherein more direct and violent kinds of politics have been played out, exploring issues from protest and direct action to political violence.

THE POLITICS OF MOBILITY

Ideology

As we saw in Chapter 2, our ideas and assumptions about mobility are really important. Mobility is frequently *ideological*, embedded within the most overt political discourses. As we saw, mobility has repeatedly been taken as a necessary bedfellow alongside our ideas about freedom or liberty. Nick Blomley (1994a) reminds us how 'mobility (of certain forms) is central to the liberal pantheon, to the extent that liberty and mobility are almost interchangeable' (Blomley 1994a: 175–176). It has been suggested that the dominancy of neoliberal politics embodies a set of assumptions undergirded by the signature of mobility. Emerging from Thatcher and Reagan's commitment to free trade, privatization and de-regulation in the 1980s, neoliberal ideology celebrates unfettered mobility for people and things. Underpinning economic policies such as the Canadian-US Free Trade Agreement (1989) neoliberal thought is embedded with an ideology of uninhibited mobility and circulation and acts to lubricate capital and people flows (see the section on citizenship below). More important, these ideas make a considerable difference to the way people are treated.

Ideological values create actions that may work to impede, constrain and disenfranchise other people as we saw briefly in the previous chapter.

We could take as a quick example the issue of car-travel once more. The assumption of car ownership in the United Kingdom has resulted in both a culture and a landscape in which mobility is both expected and necessary to participate in society. Pinioned by an ideology of universal mobility for all, the UK government's *Roads for Prosperity* (1989) white paper premised its outward planning on road building in order to supply the predicted demand in car ownership and use (Vigar 2002). Just the title of the report already signifies the alignment of automobility with notions of progress and wealth creation. But these sorts of assumptions carry with them unforeseen consequences and unequal implications. In another context, Jensen and Richardson (2004) illustrate the short-sighted ideological dominance of 'frictionless mobility' in Europe for environmental sustainability. The following case study illustrates how an ideology of automobility had catastrophic consequences for the inhabitants of New Orleans in 2005.

Case study 3.1: THE EVACUATION OF NEW ORLEANS

The tragic consequences of ideologies of universal mobility have been carefully discussed by several authors examining the impact of Hurricane Katrina on the inhabitants of New Orleans during 2005. As various writers on the events show, the evacuation plan for the city seemed to be predicated on an assumption of unified access to private transportation. This can no doubt be linked to all sorts of constructions of the American way that sees autonomous private mobility as essential to its foundation. In New Orleans an assumption of 'mobility privilege' persisted, operating 'both culturally and materially through the changes to the built environment'. In the context of New Orleans new forms of mobility infrastructure appeared dedicated to this ideology giving speedy links to the suburbs by the multi-lane Lake Pontchartrain Causeway which exacerbated further investments to these areas and to their movement pathways (Bartling 2006).

For the inner-city areas and the ethnic populations that lived there, their own mobility capacities appeared not to be taken care of in quite the same way. Although just 5 per cent of the non-Hispanic White population of New Orleans did not have access to private transport, this contrasted markedly to the 27 per cent of Black people with the same constraints.

Disinvestment in public transport systems and a reliance upon an assumption of an almost universal access to the car meant that those who didn't have access to a car would be subjected to immobility and potentially death (Bartling 2006). As Bartling explains,

> the privilege of particular forms of mobility allows people to look at the world through a lens that appears objective but in reality contributes to a wide variety of practices that ensure the continuance of particular forms of dominance.
>
> (2006: 90)

Although the evacuation of New Orleans exhibits a clear racial politics of mobility access, it is because the treatment of mobility is assumed to be de-politicized and universal, that mobility is 'emptied' of its social content and powers to socially differentiate (Cresswell 2006). 'Universalistic language' as Mimi Sheller purports, 'masks differential processes' (Sheller 2008: 258).

Further reading

(Bartling 2006; Shields and Tiessen 2006; Tiessen 2006)

The ideological associations of mobility with liberty, freedom and universalism, therefore, contain serious shortcomings. As Janet Wolff puts it, 'the consequent suggestion of free and equal mobility is itself a deception, since we don't all have the same access to the road' (Wolff in Morley 2000: 68). In other words, ideologically charged mobility politics and policies may not work because they fail to assume that mobilities are incredibly uneven and differentiated. Nomadism can suggest 'ungrounded and unbounded movement' and the resistance of 'fixed selves/viewers/subjects' (Wolff 1993). Failing to include difference and to provide for anything that compromises the norm can cause problems, hardships and in the instances discussed earlier, death.

Wolff criticizes the vocabulary of mobility as a metaphor that 'necessarily produces androcentric tendencies in theory' through the mobilization of terms like *nomad* and others. This tendency is mirrored in social life. The mapping of mobilities as a medical pathology has served to instil mobility, at various moments in time, with a morally corrupting significance. Several studies of homelessness show how the scientific pathologization of mobility led to numerous practices aimed at constricting or excluding people's movement. In the context of the United States, Tim Cresswell shows the association between homelessness and syphilis. Homeless came to insinuate 'a looseness of morals and disconnection from normality. It was mobile people such as sailors, soldiers and tramps that were seen as the spreaders and even the causes of the disease' (Cresswell 2002). Ideological inscriptions such as the pathological further upscale their significance by 'bringing into play metaphors of national health and the condition of the body politic' (Reville and Wrigley 2000: 6) as shown by Nyamnjoh in the previous chapter.

The same is also true for the treatment of women. An unwritten assumption for many feminist scholars is that 'Movement belongs to men' (2003: 3) as Scharff states with clarity. The assumption that women 'move seldom and reluctantly, and when they do it's a departure from their real stories' threatens the equal treatment of women and the provision of mobility systems for them. For example, Uteng and Cresswell note that 'Feminine mobilities are different from masculine ones.' Labelling and assuming these differences, can 'reaffirm and reproduce the power relations that produced these differences in the first place' (Uteng and Cresswell 2008: 3). Seeing women's mobility as unexpected and a 'surprise', as we will see later, has impaired mobility infrastructures to adequately serve their needs.

What, moreover, are the implications? How does the impaired mobility face reduced equalities of opportunity to participate in civic life (Kenyon 2001, 2003; Kenyon *et al.* 2002, 2003)? The treatment of mobility with something like universalism – can pose serious questions of an aim like social inclusion. As Kenyon, Lyons and Rafferty put it, social inclusion may be threatened because of 'reduced accessibility to opportunities, services and social networks' which may be due to 'insufficient mobility in a society and environment built around the assumption of high mobility' (Kenyon *et al.* 2002: 210–211).

Participation and civil society

Mobility affects whether people can truly participate – whether they can access the spaces of public deliberation and whether one's voice may be heard in a public forum. How one is mobile and the kinds of quality of their mobility are seen as a key component in processes of social inclusion and exclusion. The processes through which 'people are prevented from participating in the economic, political and social life of the community' (Kenyon *et al.* 2001: 210–211) have been explored and sought out.

On the other hand, mobility is regularly viewed as a force not necessarily for good. We should see this in relation to arguments presented earlier concerning the destruction of meaningful senses of place (Chapter 2). Alvin Toffler (1970) goes even so far as to argue that the ephemerality of the mobile world has created a 'temporariness in the structure of both public and personal value systems' (cited in Harvey 1989) and, as such, provides no grounds for consensus and common values. Mobilities not only erode our sense of place but they also destroy the spaces where one could become *public* and, therefore, political by communicating one's voice into the social field. Mobilities are seen to disrupt natural divisions such as private/public or home/away (Norton 2008).

Turning to philosopher Jurgen Habermas, Mimi Sheller and John Urry (2000) highlight how divisions between the private and public sphere are not so much connected by mobilities but completely obliterated by them. The public spaces of the city and the private places of the home appear to have been 'drowned-out in the modern urban built environment centred on traffic flow'. The space of the car, particularly seems to have replaced meeting spaces, places of debate, face-to-face interaction and, importantly, meaningful engagement. The car is suggested to have altered the role and function of public spaces such as streets and squares to the degree that they have become subservient to the 'technical requirements of traffic flow'. The result of all of this, as Habermas suggests, is that the 'space for public contacts and communications that could bring private people together to form a public' (Habermas cited in Urry and Sheller, 2000: 742) are simply erased.

Taking a historical point of view, J. B. Jackson (1984) writes how the growth of mobility infrastructures gradually erodes participation because of its tendency to support some kinds of mobility, along with the people who have access to them, while degrading others. The first highways, Jackson writes, 'made it easy only for a certain class in society to come together: officials and political, religious, and military leaders: they could meet and transact public business at selected centers' (see also Sennett 1970, 1998: 37). Mobility failed to emancipate because it instead fostered the unequal positioning of the already marginalized, 'the rank and file, particularly those in the countryside, were doomed to immobility and to political inaction' (Jackson 1984: 37). Chella Rajan (2006) writes how the removal of the pedestrian from urban space sees the amputation of the citizen member of civil society. Similarly, Richard Sennett (1990) has noted the relationship between the death of street life and the rise of individualism. Automobilization effectively removes the 'streets' from the pedestrian activists' grasp where 'the right to dissent', for Rajan, 'is then defined in terms of a prescribed set of subjects allowed to express controlled forms of "no!" within an automobilized society' (2006: 126).

But perhaps we shouldn't take such assumptions unquestioned. How can mobilities and automobilities actually permit connections to be made between the private and public sphere in ways that pull civil society into the spaces of one's home? How is 'the auto-freedom of movement' just as much a 'part of what can constitute democratic life' (2000: 742–743) as Sheller and Urry ask? We should be reminded again of Jackson's (1984) earlier writings on the highway, which, he argued, has the dichotomous capacity to not only 'bring people together' and create a 'public place for face-to-face interaction and discussion' but also push people apart. Naturally all of this is dependent upon one's ability to move and to use these spaces. And just as we tried to mobilize our senses of place earlier, the positioning of mobility as the ultimate enemy to democratic life seems to cast civil society in especially fixed and bounded terms. Just as scholarship has been able to recapture a more mobile sense of place (Massey 1993), seeing civil society as an irrevocably rigid and closed-off structure serves to make mobility its antithesis. Quite a different comprehension of civil society could place mobility in a very different way to its workings. Sheller and Urry (2000) imagine civil societies as rather like 'sets of mobilities flowing over' spaces and infrastructures such as roads. It is conceived less like a site or a place but a node or a 'crossroads'.

With reference to more mediated forms of mobility and communication, we can catch a glimpse at how these physical 'crossroads' are being replaced by other sorts of virtual mobilities. How can we think about the enablement and connection of new public and private `spaces' in the context of mobile communication and `splintered urbanism'? Sheller asks (2004: 42). Similarly, others argue that it is through virtual mobility that people can gain a *voice*, articulate their opinions and take part in or participate in discussion and debate without the need for physical co-presence, thereby 'overcoming some of the temporal, confidence, financial and mobility constraints' (Kenyon *et al.* 2002: 214).

According to politics theorist Chris Rumford (2006), mobilities, therefore, offer a new and changing 'spatiality of politics', witnessed and represented by spaces that are far less willing to conform to the territorial spatiality of the nation-state, or a boxed-off idea of civil society. These are new 'public spheres, cosmopolitan communities, global civil societies, non-proximate or virtual communities, and transnational or global networks' (Rumford 2004: 161).

Power geometries and the politics of difference

Our final and substantial address of the politics of mobility looks closer at the association of mobility with liberal assumptions of social justice. A key proponent of these ideas has been once again geographer and social theorist Doreen Massey. As introduced in Chapter 2, Massey has been influential in re-theorizing mobility away from its nomadic, romanticized, and reactionary interpretations and I refer you to the following key ideas box for more detail on the evolution of Massey's thought.

Key idea 3.1: THE POLITICS OF MOBILITY AND POWER GEOMETRIES

Doreen Massey's formulation of a 'power geometry' is essential to our understanding of the relationship between mobility, politics and especially power. Massey took particular issue with David Harvey's (1989) use of terms such as 'time–space compression' (see Chapter 5). Running with the idea as a way to signify the effects of increased global mobility, Harvey treated mobility as an unproblematic term. 'Time–space compression has not been happening for everyone in all spheres of activity,' Massey summarized in her critique (1993: 60). The point Massey was searching for was, 'where is difference?' And, furthermore, how could a politics of mobility identify the key moments and instances of exclusion and marginality.

These issues were captured in several examples of how different people had quite different sorts of experiences of mobility. Tracing an ironic juxtaposition between the mobile rich and the mobile poor, Birkett's work on the Pacific Ocean provides a compelling material example:

> Jumbos have enabled Korean computer consultants to fly to Silicon Valley as if popping next door, and Singaporean entrepreneurs to reach Seattle in a day. The borders of the world's greatest ocean have been joined as never before. And Boeing has brought these people together. But what about those they fly over, on their islands five miles below? How has the mighty 747 brought them greater communion with those whose shores are washed by the same water? It hasn't, of course.
>
> (Massey 1994: 148)

Massey is making what is a very simple yet vitally important point. While mobility has brought time–space compression to those who can afford it, many people cannot experience its benefits so acutely because they simply do not have access to it. Thus she states, 'most broadly, time–space compression needs differentiating socially' (1994: 148).

Further reading

(Massey 1993; Cresswell 1999a, 2001; Hannam *et al.* 2006)

As discussed above, Massey is interested in developing a politics of mobility in order to explore and analyse how mobilities are socially differentiated and unevenly experienced. Massey seeks to offer alternative frameworks to universalist ideologies in order to take difference seriously while accounting for the identities and capacities that diverge from any kind of societal 'norm'. Mobilities and the process of time–space compression do not spread evenly over society as if it were an undifferentiated plain – the sort of assumption we witnessed in the previous chapter, 'where movement is *possible* in all directions' (Dicken and Lloyd 1977: 45).

There is, however, a little more to it than that. Access to mobility and certain types of mobility are less than random. While mobilities may be differentiated socially, these differentiations reflect and express already existent social differences and hierarchies. Let us read from Massey a little more:

> In a sense at the end of all the spectra are those who are both doing the moving and the communicating and who are in some way in a position of control in relation to it – the jet-setters. (1994: 149)

What Massey is suggesting is that people from both ends of an already unequal and hierarchical society will have quite different levels of access to certain kinds and qualities of mobility. Some people might be able to afford it and are in control of it, while others might not. Some might have the time for it, while others simply don't.

Positioned towards mobility in these ways, some people get to take advantage of it. The sorts of people Massey mentioned were seen to be 'groups who are really in a sense in charge of time–space compression, who can really use it and turn it to advantage, whose power and influence it very definitely increases' (Massey 1994: 149). Just as some people might have better access to mobility, once gained, they may use their mobility to reinforce and improve their social standing. Those who can afford to drive a car, for instance, or can afford a toll-road perhaps, may be given greater opportunities for employment, while, say, someone who couldn't afford it might not. The ultimate issue here then is much more than a recognition of difference; it is rather how those differences are reflected and come to reinforce societal inequalities and differences.

Mobile method 3.1: THE PRIMACY OF POSITION

Some of the most well-known investigations of the airport terminal have often been written by an academic enjoying a very specific mobile experience of the process and site of travel. The inevitable danger is that the mobile scholar may suppose that their experience of mobility is the same as anyone else's.

As Beverley Skeggs argues, many writers on mobility 'rely on a safe and secure place from which to speak and know . . . theorizing and legitimating their own experience of mobility and self, and claiming it as universal' (Skeggs 2004: 60). In this way, the very real problems of universalism discussed above become perpetuated in academic methodological practice. Geographer Crang sees how the consequent 'singular ego-ideal' is 'sutured into the image of the forty-something, healthy male businessman' often at the 'exclusion of other identities' (Crang 2002a: 571). Similarly, Peter Merriman (2004) shows how Marc Augé's popular writings on airports, motorway stations and other places of mobility have been assembled by relying upon the experiences of the business executive. This tendency generates accounts that are dependent upon the business traveller's 'frequent experiences of traversing and dwelling in these spaces' that may be interpreted as 'familiar', 'expected' and particularly routinized. This has massive implications for what we may conclude about spaces like the airport, as anthropologist Orvar Lofgren writes: 'what is an airport like Kastrup, Heathrow, Kansai? The answer still depends on what kind of traveller you are' (Lofgren 1999: 19).

Thus, according to Marcus, 'constantly mobile, recalibrating practice of positioning' is, therefore, always composed in the context of the researchers' moorings, and in their affinities and repulsions from those they do research with (1995: 113).

In practice

- Consider your own position in your ethnography and participant observation of travel
- You might think about whether it is possible for you to step into different subject positions and examine what sort of issues occur as you do

Further reading

(Augé 1995; Crang 2002a; Merriman 2004; Skeggs 2004)

Since Massey's intervention many writers, thinkers and academics have extended her work into different fields. Anthropologist Aiwah Ong's (1999) writings surrounding citizenship question the 'misleading impression that everyone can take equal advantage of mobility and modern communications' and the familiar assumption that mobility can be 'liberatory' in 'both a spatial and political sense, for all peoples' (1999: 11). Ong seeks to trace out the processes and transformations that may stimulate strategies of mobility for some, while others must endure 'staying put', or 'being stuck', in place. From *Media and Cultural Studies*, David Morley considers Massey's 'geometry' in the context of new forms of communication and mobility. For Morley (2000: 199), what is important is even more fundamental than mobility itself, that is 'who has control – both over their connectivity, and over their capacity to withdraw and disconnect'. Further, he suggests that it 'matters little whether the choice is exercised in favour of staying still or in favour of movement' (Morley 2000: 199). Similarly sociologist Zygmunt Bauman's (1998) treatise on *Globalization* reveals a society divided by the way it moves. Bauman conceives society as a filtering 'difference machine' that sorts and distributes opportunities to move unevenly across its population.

Reminiscent of Harvey's arguments explored earlier in the book, Bauman shows how it is through the globalizing processes of international business and finance as well as information flow and the paradoxical strategies of space fixation, that penetrating differentials of movement are revealed. For those 'high up' in the social hierarchy, mobility comes with ease while for those 'low down' their choice to be mobile is far more restricted. Even though some may enjoy being fully 'global', others are well and truly fixed in their 'locality'. As Bauman puts it, 'Like all other known societies, the postmodern, consumer society is a stratified one' (1998: 86). The way in which people move in these societies is, therefore, one of the greatest indicators of one's relative position in society (see also Bourdieu discussed in Chapter 4). Imagining a histogram-like chart representing society, the 'high-up' and the 'low-down' classes are plotted against their *degrees of mobility* – their 'freedom to choose where to be' (Bauman 1998: 86).

Using these kinds of hierarchies of verticality to describe divisions of 'high' and 'low' is incredibly suggestive in more than a metaphorical sense but in the sorts of realities they speak to. Discussed with reference to several examples in the case study box below, those higher up in the hierarchies of class may come to enjoy literally spaces higher up above the ground. Verticality can often mean speed, too, as we find a complex politics of speed and verticality combining to fix unequal hierarchies of connection and disconnection.

Case study 3.2: THE STRATIFICATION OF HELICOPTER TRAVEL IN SÃO PAULO

Academic studies seeking to examine the vertical dimensions of urban mobility networks have shown how vertical stratification, speed as well as mobile issues of access and quality are tied up in unlikely ways. One of the most sustained critiques of this predominantly urban phenomenon can be drawn from Graham and Marvin's *Splintering Urbanism* (2001) which follows Mike Davis' (1990) critical exploration of 'skywalk' structures of Los Angeles that are divorced from the local street network. These building developments, found in other North American cities and increasingly urban areas in South Asia, offer well-connected office complexes and shopping malls that bypass local street networks. The effect is to separate from and filter out undesirables on the ground below, creating a citadel or fortress-like space (Davis 1990) whose inhabitants can move quickly between. For Boris Brorman Jensen, this sees 'the ideal of integration replaced by a practice of differentiation' (2004: 202).

Height and verticality mean and signify disconnection for others, while improved connectability and speed for someone else. By bypassing local networks of mobility, places farther off become much more accessible. David Morley quotes from Mark Kingwell who, in watching helicopters take off from distant office buildings, notices how verticality often means the privilege of speed: '[E]xtreme speeds are not available to most of us,' (cited in Morley 2000: 199).

In cities such as Bangkok, Thailand, dense networks of streets and roads are similarly differentiated and vertically 'dispersed' according to a 'hierarchy of speed' (Jensen 2004: 186). Private consortiums of road builders have built expensive tolled expressways in order to avoid the constipated periods of rush hour. The elevated 'sky train' (a ticket costs ten times the price of an ordinary bus ticket) speedily connects business districts, tourist centres and hotels, allowing passengers to pass above the invisible, densely populated, poorer areas below. Once they step off the train, they can glide straight into the adjoining shopping mall or hotel complex (Jensen 2004).

Taking the subject of vertical travel further, Saulo Cwerner's (2006) article on the private helicopter user market in São Paulo, Brazil draws out some of Graham and Marvin's initial observations in more detail. Cwerner explains how São Paulo's incredible growth in helicopter traffic reflects and enables a marked segmentation of mobility use in the city. For those unable to afford it, constant gridlock reaching lengths of 150 to 200 km is an everyday experience due to 5.4 million cars that occupy the city. Helicopters allow their wealthy owners or users to bypass this congestion so that they may reach other office blocks within the city with ease or travel to their homes in the suburbs. Resembling the gated communities in which their users may live, helicopter travel permits a gated and secure tunnel of movement that joins up with the security afforded to their residence (see figures below). Fear of violent crime and particularly kidnappings, Cwerner argues, are one of the primary impetuses for the rise in helicopter use. In São Paulo, a city in which homicides occur six times more regularly than in New York, 'the mobility imperatives of the economy of flows goes hand in hand with a perception of chaos spreading across the city below' (Cwerner 2006: 203).

Further reading

(Cwerner 2006; Davis 1990; Graham and Marvin 2001; Jensen 2004)

Conceptualizing mobilities in this way conjures up an already differentiated society that re-maps itself over mobility. In this case previously unequal social relations reproduce themselves through mobility and thus reinforce their differences. People may have little choice because of the sorts of mobile societies they must move within; how one is placed within the rest of society effectively constrains their opportunities to be mobile.

Tracing the lines and the limits of an already differentiated economy and society has led some authors to theorize how these lines form a boundary, or a structure of sorts that sets the limits of one's capacity to move. Remember the pioneering work of Torsten Hägerstrand (1982) discussed earlier, Hägerstrand found that people's movement patterns or time–space routines were really rather more solid and stable than at first thought. Even though the patterns he recovered were the outcomes of behaviour, they acted as structuring constraints upon people's movements, being a wider reflection of social networks of friends and communities. It was argued that people had little ability to actually change this structure very much being born into it.

Various fields have utilized these ideas more than others. Take the case of tourism studies which deploys the notion of a 'space–time prism'.

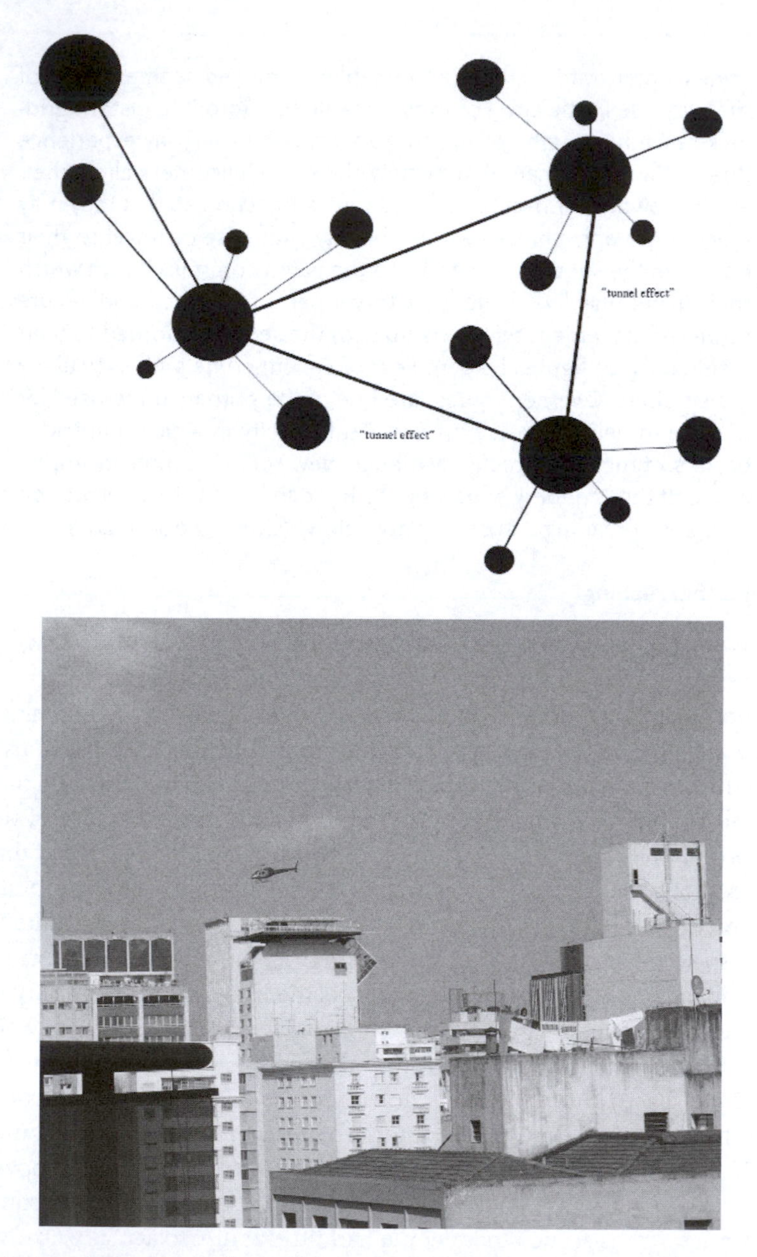

Figure 3.1 The 'tunnel effects' of 'hub and spoke' infrastructural networks and elite helicopter travel in São Paulo

Source: After Graham and Marvin (2001, orig. fig. 5.1) and copyright © Saulo Cwerner, respectively

In this formulation people's potential mobilities are governed and constrained by the particular space–time prism they belong to, their prism being defined by their class, identity, income or other characteristics. Closely resembling the Hägerstrand model, C. Michael Hall (2005) articulates how this prism might work. He explains that certain kinds of monetary and temporal capital are required in order to pay for and fulfil certain mobility projects. Cash- and time-rich people consequently have much more potential to move than those without. Like Hägerstrand and others before him, Hall points out that it is just as impossible to break or surpass this structure, so much so that he argues, 'In one sense the prism can almost be referreed to as a prison as it is impossible to move beyond it' (Hall 2005: 79). Clearly, there must be undoubtedly much more to mobilities than being cash or time rich and they don't necessarily lead to each other. People very cash rich may have little time at all. Consider the impact of commuting once more. Those cash or time rich may experience many other factors that stop them from being mobile.

Nonetheless, Hall's attempt to unify spatial mobility with that of social mobility nicely represents the social-spatial structures, position or limits, which may not determine but certainly influence one's ability to be mobile. Although other authors to lead this work have overplayed the apparent loss of social context within examinations of spatial mobility (especially if one reads the work of Cresswell 1993, 1996; Thrift 1996), they suggest one other important component to their critique: mobility is too often taken as an actual or a past thing. In other words, what are the factors involved in how mobility could occur – how opportunity or potential actually gets turned into mobility? One might ask the question about the importance of this potential too. What significance does potential to move hold? How is it dealt with in a social way? In asking these questions several re-theorizations of potential mobility or mobility opportunities have evolved, currently orbiting around the idea of *motility* (discussed in the key ideas box below).

Key idea 3.2: MOTILITY AND MOBILITIES AS CAPITAL

Vincent Kaufmann and his colleagues Manfred Max Bergman and Dominque Joye (2004) are key contributors to the sociological application of 'motility', a concept previously found within the biological sciences (see Bauman (2000) for an exception).

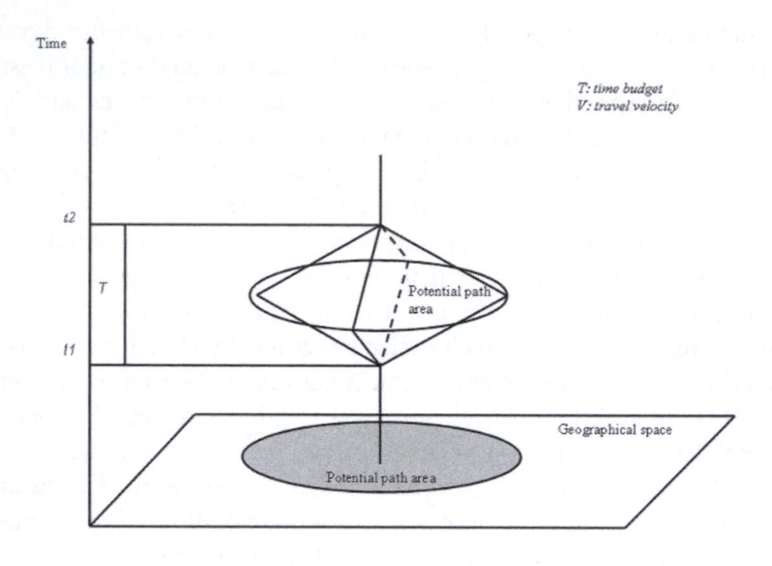

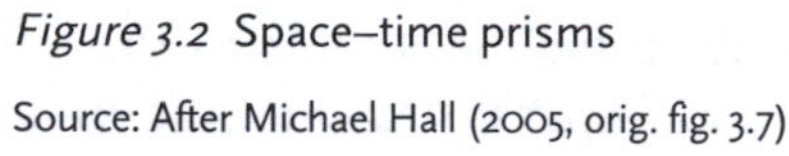

Figure 3.2 Space–time prisms

Source: After Michael Hall (2005, orig. fig. 3.7)

In his book *Rethinking Mobility* (2002), Kaufmann describes motility as simply 'the capacity of a person to be mobile', or as he puts it more precisely, 'the *way in which an individual appropriates what is possible in the domain of mobility and puts this potential to use for his or her activities*' (Kaufmann 2002: 37). Kauffmann's contextual approach to mobility sees social and spatial mobility as interdependent modalities. One's capacity to be mobile is dependent upon all sorts of social, political, cultural and economic contextual variables from 'physical aptitude', 'aspirations to settle down', other 'existing technological transport and telecommunications systems and their accessibility' to 'space–time constraints' such as the 'location of the workplace'. Motility is not only much more than the potential to be mobile but is also the ability to turn that potential into an actuality. 'Access' to mobilities, 'skills' or the competency of a person to make use of this access and the 'appropriation' – how these variables are evaluated and transformed into mobility – appear as the key factors for mobility usually involving some kind of compromise between them.

The final and perhaps most important strand is how they ally motility or movement with 'capital'; they propose how motility 'forms theoretical and empirical links with, and can be exchanged for, other types of capital' (Kaufmann *et al.* 2004). Motility is thus commensurable and, therefore, exchangeable as a sort of commodity. This helps to flatten out epistemological differences between physical movement and its potential with monetary capital. It allows the imagination of how our potential mobilities can lead to or be exchanged for other sorts of financial or social capital. Take Kauffmann's description of a woman aspiring to a professional career in advertising, who has acquired considerable motility by her career choices and by learning two foreign languages. At the same time she is also moored by her marriage to her husband whose business is well established, and they plan to have children and to own a home (Kaufmann 2002: 45).

Analysing the above through motility leads to an extraordinary interchangeability, a capacity to exchange motility, mobility and the potential accumulation of wealth. The woman's potential mobility and willingness to migrate appear to raise the chances – or the potential of her achieving her career goals. On the other hand, her rootedness to her relatively immobile partner and the possibility of being tied to a family home, work in the other direction to decrease her mobility potential.

Further reading

(Kaufmann 2002; Kaufmann *et al.* 2004; Kesselring 2006)

Motility and its transformation into mobility seem to require some kind of negotiation or management. What Kaufmann describes as a compromise suggests a sort of balancing between one's aspirations, lifestyle and personal characteristics. This is vitally important in recognizing the limitations of what Kesselring calls 'autonomous mobility politics'. Freedom of movement and autonomous decision making do not take place automatically, as they are rather a process of management that involves 'juggling and struggling with mobility constraints'. Thus, people work out strategies in relation to their working lives and constraints, to the restrictions and dynamics of transportation, for instance, and many more (Kesselring 2006: 270–272). Understanding mobility as an accomplishment of motility into

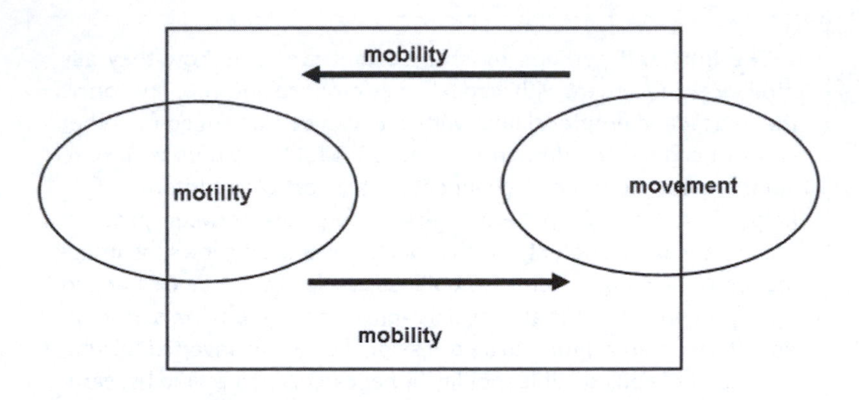

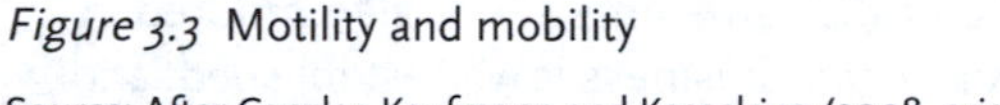
Figure 3.3 Motility and mobility

Source: After Canzler, Kaufmann and Kesselring (2008, orig. fig. 1.1)

movement lies at the heart of the most recent conceptualizations (Canzler *et al.* 2008) (see Figure 3.3 above).

Motility moves us into the final stage of this geometry in the way it demands that we consider how these prisms, societal differentiations and mobility constraints are intimately related to one another in the way that they have effects and exert force. There is considerable tension along the geometry as these forces push, pull and flex producing unequal and uneven consequences. Hannam *et al.* explain how investigating 'mobilities thus involves examining many *consequences* for different peoples and places located in what we might call the fast and slow lanes of social life' (2006: 12, emphasis added). A consideration of the consequences of mobilities demands that we pay attention to the other side to Massey's power geometry. Hannam *et al.* are interested in the 'places', 'technologies' and 'gates' 'that enhance the mobilities of some while reinforcing the immobilities, or demobilization, of others' (Hannam, Sheller and Urry 2006: 12). Similarly Skeggs writes how it is necessary 'to understand who *can* move and who *cannot*, and what the mobile/fixed bodies require as resources to gain access to different spaces' (Skeggs 2004: 48).

There is a flipside to this geometry which we can see by twisting it round to examine more than how one is placed in relation to it – and in relation to others – but how the geometry produces uneven effects. We need to think about not only how one is placed in relation to it quite differently but also how one is *affected by* mobility quite differently. We can see this working once more through Massey's contemplation on the island of Pitcairn. In juxtaposing the increasing affordability of air travel and the concurrent decline in shipping, Birkett shows how the mobility of one group can impact another's, so much so that the 'The 747s that fly computer scientists across the Pacific are part of the reason for the greater isolation today of the island of Pitcairn . . .' (1994: 148).

This may feel like a rather abstract example so Massey asks the reader to consider the implications of a transport technology like the private car. What is the effect? Upon whom does one's mobility impede? Massey suggests that while increasing personal mobility, 'they reduce both the social rationale and the financial viability of the public transport system – and thereby also potentially reduce the mobility of those who rely on that system' (1994: 150). The example of the out-of-town shopping centre is extremely useful here, if we consider how moving to it and relying upon it as a regular destination to buy our shopping contributes to the failing viability of the corner shop, the smaller retailer and potential transport services into town and village centres.

Clearly, the issues are much more complicated than this and connote far more plural and complex politics if one considers the example's relation to other forms of transportation such as the bike; whether taking the car would reduce pressure on an already crowded bus service – which could have provided a barrier to the mobility impaired or a mother laden down with shopping and children; or the public and private subsidies that may or may not have supported that particular bus service. Having set up apparently indissolvable divisions and hierarchies, we should remember that they are not so simply constructed. We have seen how toll roads and premium-space highways create inequity between those who use them and those who cannot. On the other hand, they can have more equitable implications. Road congestion charging in London and the taxing of public car ownership in Singapore has had the effect of generating investment monies in respective public transportation systems (Livingstone 2004; Wolmar 2004; Santos 2005). But even then, congestion charging schemes may be rolled out at the expense of the most 'at-risk' groups. Bonsall and Kelly remind us that 'Not all car owners are affluent' (Bonsall and Kelly 2005: 407). Taking premium network-mobility spaces on their own, and not in relation to the wider mobility systems they interact with, can sometimes hide their more extensive and positive implications.

In the case of Wiggens, her mobility was entirely dependent upon the mobilities of customers probably travelling by car to the shopping mall at which she worked. Customers would no doubt move to the mall with the reasonable expectation that Wiggens and her co-workers would be there to service the various retail outlets they had come to visit. At the same time, these very mobilities provided the greatest barrier to Wiggens' access to the mall. The highway and the busy traffic of cars and trucks were felt in much more serious ways than the experiences represented by Baudelaire as discussed earlier. Customers travelling to the mall by bus would not be dependent upon the cars, but they would be just as much affected by them and the mobility infrastructure that supported them.

Following a central premise of this book, looking at mobilities through this kind of geometry means you can never take mobility on its own, ever. Mobilities always have relational impacts and we must question what those are. For Massey this is a question of power exerted perhaps intentionally or unintentionally with predetermined or indeterminable consequences. We must ask who and what other mobilities do our mobilities effect? As Massey puts it, we must question 'whether our relative mobility and power over mobility and communication entrenches the spatial imprisonment of other groups' (Massey 1991: 151).

Cresswell's referral to the 'symbiosis' of mobilities (shown in Chapter 1) cements Massey's notion, for just as some mobilities could be dependent on the kinds of immobilities or moorings, to use Urry's terms, 'one mobility may be symbiotically related to other mobilities with entirely different cultural and social characteristics' (Cresswell 2001: 21). This power play can take place within the complex relations between mobilities, economies and public transport investment we have seen.

For Massey, 'it does seem that mobility, and control over mobility, both reflects and reinforces power' (1991: 150). But it is more than a question of how some have more mobility than other people, that mobility is dished out unfairly. The question is more that, 'the mobility and control of some groups can actively weaken other people. Differential mobility can weaken the leverage of the already weak' and thereby undermine the power of others (Massey 1991: 150).

ENTANGLEMENTS OF MOBILITY

These three dimensions of mobility: *ideology*, *participation* and their *geometries of power* are tangled up in very knotty ways. Ideologies can play an important role in various policies towards mobility and ways of treating it. These exertions may well serve to differentiate mobilities, perhaps unfairly so. And the consequences of these relationships may constrain one's ability to find employment, reach essential services or participate in a public sphere. What we are articulating, therefore, is a kind of politics that is about questioning and assessing unquestioned conceptions of mobility and freedom, gathering alternative frameworks that may develop 'a recognition of difference and responsiveness to individuated needs, as well as the protection of the rights of difference' (Imrie 2000: 1653).

The chapter now realizes several examples of these ideas through a set of research themes and case studies. The first is citizenship.

Citizenship

One of the most obvious and well-studied aspects of the politics of mobility is concerned with citizenship. Citizenship is one of the most complex issues we could deal with, yet its intricacy is often hidden within the simple object of something like the passport. As a pertinent symbol and exemplar of the intertwinement of mobility and citizenship, political historian John Torpey's (2000) *Invention of the Passport*, and similarly international relations scholar Mark Salter's (2003) *The Passport in International Relations*, sees it as something much more complex. The passport as well as the paper documents that came before them have been issued to citizens by the nation-state in order that they could manage and facilitate travel – to 'monopolize the legitimate means of movement' (Torpey 2000) – both within and across borders. Citizenship involves the negotiation and management of mobility (Hindess 2002; Walters 2002b,a; 2006), and the passport is both a tool to monitor and manage it by the state, while it allows citizens to prove their identity and claim the rights of travel.

We shouldn't overlook that modern paper documents and passports have come long after citizens of liberal democratic societies and nation-states were endowed with rights to move. The right to be mobile was even written into the Magna Carta. And although (Cresswell 2006b) these rights were not actually present within the constitution of the United States, through various court cases it was debated and deliberated as a fundamental part of what it meant to be a citizen of the United States (in the context of Canada see Blomley 1992; Blomley 1994b). Mobility is held within 'longstanding ideas of the nation and what it is to be a citizen, as well as activity that constitutes "commerce"' (Cresswell 2006b: 750).

A wealth of writers have begun questioning how the logics of mobility and citizenship have altered in an era of neoliberal ideological dominance which has seen policies, international agreements and treaties alter the dimensions and surfaces of national borders in order to facilitate capital flows and personal cosmopolitanism (Tomlinson 1999; Beck 2000). In considering an innovation in border control technologies, Matthew Sparke (2004) focuses on a speech made by US Department for Homeland Security Director Tom Ridge in advance of the Smart Border Declaration agreement between the United States and Canada in 2001. Ridge's prophetic alignment of the passport with the credit card gestures towards increasingly consumer-driven (what Craig Calhoun [2002] calls consumerist cosmopolitanism) and economically led forms of post-national citizenship.

According to Ong (1999, 2006), and explored in more detail in the key ideas box below, new forms of flexible citizenship have emerged in response to the liquidities of capital and more permeable borders. Flexible citizenship refers to what Ong calls 'the cultural logics of capitalist accumulation, travel, and displacement' which have encouraged or 'induced' people to 'respond fluidly and opportunistically to changing political-economic conditions' (1999: 6). Here, the mobile businessmen of Hong Kong, many of whom hold multiple passports as an insurance against Chinese–Communist state rule, are an exemplar of an increasing body of 'mobile subjects who plot and manoeuvre in relation to capital flows' (1999: 6).

Key idea 3.3: FLEXIBLE CITIZENSHIP

Aiwah Ong's multiple accounts of migration and citizenship point to a new logic of flexibility in people's movements across the globe and how they come to belong to particular places. While these at first may appear like an escape from the State, Ong later renders 'flexible citizenship' as a new 'mobile calculative technique of governing' (Ong 2005: 13).

Assessing the movements of Chinese businessmen to the United States and Canada, Ong examines how this kind of multiple or flexible belonging implies continuous flexibility in one's social and geographical positioning. Their careful choice of investments, work and family relocation require similarly careful negotiations between various layers of their lives, their family, the state, and capital.

Enabling these mobilities are state regimes which are able to respond to and attract productive mobile capital and labour. Immigration laws have allowed many mobile businessmen to operate between the Pacific coasts of China and Canada having relocated their families to North America.

Ong also tracks similar agreements emerging from the United States in the 1990s, when a new 'investor category' of immigration law was created in order to attract the wealthy Chinese newcomers from Canada and Australia. A green card could be the business person's return for capital investment to the tune of 1 million dollars.

Further reading

(Hyndman 1997, 2000; Ong 1999, 2006; Roberts *et al.* 2003)

For Ong, flexible citizenship is not about the individual's escape from the nation-state, but how the state has allowed the emergence of these relations. Through new forms of state machinery and technologies of regulation, from visa agreements (Salter 2004, 2006; Neumayer 2006) to the emergence of post-national citizenship regimes (Mitchell 2001), various state-led systems have been constructed in order to encourage and facilitate the speedy international mobilities of people and capital. These provide the possibilities of *cosmopolitanism* – a term usually levelled to mean something like a 'citizen of the world' where one belongs to some kind of global civil society (Beck 2006). This sense of belonging and membership means that cosmopolitans can belong as citizens of 'their immediate political communities, and of the wider regional and global networks which impacted upon their lives' (Held 1995: 233).

By differentiating these mobilities, Bauman realizes how the sorts of citizenship regimes already discussed could be taken 'as the metaphor for the new, emergent, stratification', laying bare the fact 'that it is now the "access to global mobility" which has been raised to the topmost rank among the stratifying factors' (Bauman 1998: 87). Similarly Ong's later arguments are suggestive of who may find themselves outside of these entitlements. These may be citizens 'who are deemed too complacent or lacking in neoliberal potential' and thus 'treated as less-worthy subjects', fragmenting 'what we long assumed to be a homogenous collectivity and a unified space of citizenship' (Ong 2005: 16). On the other hand, refugees seeking asylum might make claims to their rights as global citizens (Lui 2004).

Many authors have begun to show how it is at the zones and spaces of the border crossing that these differences are both highlighted and performed as citizenship and rights may be scrutinized, accepted or declined. At the border, one's visa and passport are identifiers that identify themselves and their membership – entry is *mediated*. And yet, as Salter shows, it is at this moment of border crossing that the sovereign is able to exclude and sets the limits of its populations. Mobility is, therefore, 'structured in terms of entry, which is made obligatory by citizenship or refugee status, or entirely the discretionary by noncitizenship' (2004: 175).

The mobile point of the border becomes a zone of exception and indistinction as citizens find themselves between states and between their normal rights (Agamben 2005). Different kinds of scrutiny, detention and detainment may be performed by police and border control agents in these spaces, belonging and rights of entry are given and not necessarily taken (Salter 2007, 2008).

William Walters alludes to the way borders act less like 'iron curtains or Magino Lines, but more like *firewalls* differentiating the good and the bad' (2004: 197, emphasis added). Walters' point is not just about the informationalization of border zones but the sorting practices they enact. A separation between 'the useful and the dangerous, the licit and the illicit' is made in order to form a *clean* and *secure* territory. Take the Canadian–US flexible citizenship programme Nexus, that seeks to create economic integration in the elusive Cascadian corridor that stretches from Vancouver, through Seattle to Portland, Oregon. Matt Sparke (2005) asks what kinds of politics of difference are evident within the micro-spaces of the border zone. Who is economically valued? Who is high risk? Who has been given the membership of a post-national system and who has not?

In this differential bordering (van Houtum and van Naerssen 2002) we find Massey's geometry of power evident once more as it appears that the contradictions of new kinds of post-national border regimes revolve around both the facilitation and negation of mobilities that are entirely linked – what Heyman and Cunningham refer to as processes of mobility and enclosure (2004). Suggesting that border crossing is far more complicated than a post-national celebration, Cunningham and Heyman (2004) argue how borders have become the object of a paradoxical (im)mobility – how movement is enabled and induced – of people, goods and ideas – and how that mobility may be 'delimited and restricted' (Cunningham and Heyman, 2004, a process discussed in more detail below).

In the context of Europe, mobility is enshrined into its very idea (Jensen and Richardson 2004). Yet cultural and political theorist Ginette Verstraete (2001) discusses the tension involved in granting a 'freedom of mobility for some (citizens, tourists, business people)' that can 'only be made possible through the organized exclusion of others forced to move around as illegal "aliens," migrants, or refugees' (Verstraete 2001: 29). Completing what Walters referred to as the immobilization and removal of 'risky elements so as to speed the circulation of the rest' (Walters 2004: 197) back to the US–Canadian border, Matt Sparke suggests that the border is truly bifurcated in a binary logic of 'primary' and 'secondary' border crossers, or 'good guys' and 'bad guys'. Those put into the secondary category have to wait for longer questions, are asked for further documentation and may even have their cars and selves searched, 'they can expect to have their border-crossing considerably delayed, if not halted altogether. Meanwhile, the INS chief underlined, the service focuses much of its energies on speeding up the crossings of those in "primary"' (2005: 159).

Away from the spaces and enclosures of the border we witness the 'detachment of entitlements from political membership and national territory' (Nyamnjoh 2006: 16), as these rights and benefits are targeted towards certain people, while they are denied or taken away from others. Nyamnjoh's compelling study of citizenship in Southern Africa, as discussed in Chapter 2, illustrates such a trend as highly skilled and valued migrants enjoy even more benefits than the citizens of the communities they move into, and many more benefits than the temporary citizens who are just as mobile – those moving for domestic worker positions who are not given the same benefits and are often subjected to discrimination and subjugation.

Likewise, the political obligations of belonging to a particular place can be demanding and constraining, so much so that the lucky members who have opted for post-national citizenship may choose not to belong to anywhere at all. The personal autonomy held by these members to disengage from such obligations can lead to local distrust and even hate for the footloose mobilities of these quasi-citizens. As Hannerz suggests, 'the surrender is of course only conditional. The cosmopolitan may embrace the alien culture, but he does not become committed to it. All the time he knows where the exit is' (Hannerz 1990: 240).

Atomized individuals and the gendered transport exclusion

We can look to the sphere of transport mobilities to uncover other geometries of difference and consequences. Here we witness how ideologies of universal flow and dependency – which appear to contain a remarkable resemblance to the kinds of assumptions of paths and atoms witnessed in Chapter 2 – make a difference to people's abilities to move in public transport systems. There are several issues to dwell upon here that revolve around how the assumption of a 'universal disembodied subject' presents a figure of an able-bodied and genderless individual who creates serious implications for subjects who depart from such an imagining. As Rob Imrie puts it, these imaginings may 'serve to alienate impaired bodies and to prioritize the movement of what might be termed "the mobile body"' (Imrie 2000: 1).

Technology theorist Judy Wacjman (1991) argues how the simple patterns and routines of men and women are really a world apart. By tracing very different 'patterns of time, space and movement' Wajcman shows how contemporary urban spaces appear to be premised upon 'a mode of transport that reflects and is organized around men's interests, activities and desires' with consequences which often work 'to the detriment of women' (1991: 126). In general, women have traditionally moved in different rhythms to men, for various reasons to do with childcare, employment and social routines; therefore, in general, women's journeys are often shorter in terms of distance and the time taken to travel. They are more frequent too, occurring at different times of the day from the morning and evening rush hour. Wajcman argues that these differences in mobility-propensity marginalize women from using public transport systems because they do not cater for this difference. As Dolores Hayden states, it is about the question of 'If the simple male journey from home to job is the one planned for, and the complex female journey from home to day care to job is the one ignored' (Hayden 1984: 152).

Case study 3.3: WAITING FOR THE BUS

'[T]he bus is a city of women' writes Sikivu Hutchinson (2000) in her portrayal of the class, racial and gender politics of mobility in the context of the Los Angeles public bus system. Sikivu's starting point is the body, and how the inequalities of the city impose itself on that terrain by enforcing racial and gender hierarchies. Of course, the bus has long remained a site of racial politics, highlighted by Rosa Park's refusal to give up her seat for a White woman in 1956. For Hutchinson, the bus does more than highlight the politics of mobility inequalities but integrates into a wider infrastructural system of capital investment, route networks of roads, toll roads and trains directed towards particular identities and consumers.

Travelling across the LA sprawl, Hutchinson notes how the bus is the conveyance of 'the raced body, the transient, the low-income, the immigrant'. Her travelling companions are predominantly working-class women of colour whose elliptical movements pass through the city to their workplaces, public agencies and friends or family by way of the bus forming a distinctive cultural and kinetic rhythm. Situated in the context of the Los Angeles' strike of 1994, we see the implication for certain kinds of mobility and certain kinds of mobile people disenfranchised by the enhanced provision and investment in different kinds of mobility infrastructures. In LA this came to a head with the city's turnabout policy to heavily invest in a light railway system after years of neglect in favour of the private motor car. A bus worker's strike culminated in a legal suit charging the LA authorities with creating a two-tier transport system which discriminated against predominantly 'low-income' 'minority' bus riders in favour of white-collar rail commuters. The resulting victory charged the Mass Transit Authority (MTA) with upgrading the bus service and capping fares. As Hutchinson (2000) writes, 'The union's insistence that "improving the transit system is a civil rights issue because most commuters are minorities and have low incomes" goes to the heart of how denial of transit access, attendant to the increasing privatization of public space, "others" communities of colour' (111–112).

Further reading

(Hutchinson 2000)

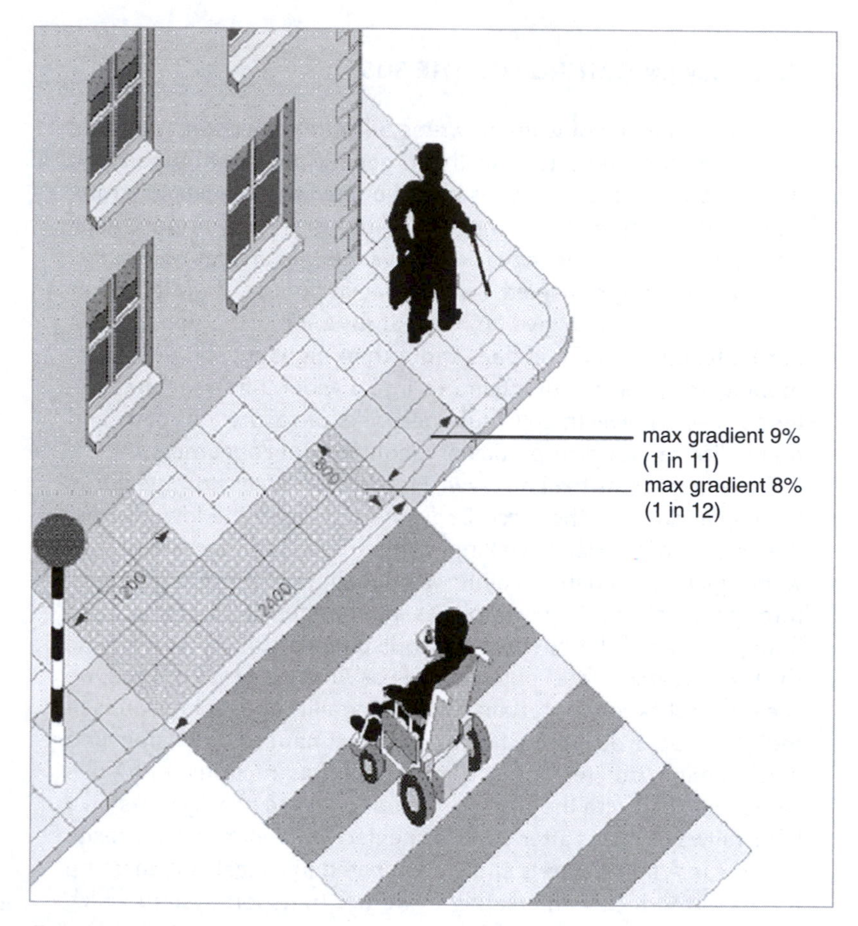

Figure 3.4 Inclusive mobility design

Source: Copyright © Department for Transport (UK) 2005

While the style and rhythm of mobilities may differ from the norm, the ease of which these lines and nodes are negotiated may vary from the imaginary model too. Envisaging an able-bodied person able to navigate and move their body through space with ease, transportation infrastructures have failed to take account of bodies which are not so adept at doing this (Oliver 1990). Understood through the social model of disability, spaces socially and materially construct disability by failing to enable their user's *different* needs. By accounting for different mobility needs, inclusive design can serve to *enable* rather than disable users (see Figure 3.4).

But just as the mobile billiard ball models render a body that is very able the atomization of body subjects is often taken far too literally. As these ideas are re-inscribed back upon society Robin Law (1999) makes the point that conventional transportation models impose their atomized individual upon subjects who are far more dependent – or far less atomistic(!) – than they are given credited for. As Law argues, 'Since women make many more trips as escorts for dependent people, the billiard ball metaphor of an independent travelling body is much less appropriate for them than for men' (1999: 582). Researchers have investigated the path dependency, especially of women, with other people. Women may be responsible for helping elderly relatives to the supermarket or unable to leave their young children at home while they go to the shops. The individual billiard-ball-style model is far from a perfect fit as many mobilities are entirely dependent upon others to move.

Forgetting these dependencies within public transport provision creates numerous difficulties. Taking children on the bus can be 'arduous', 'burdensome' and a 'trauma', travelling back from a supermarket with heavy bags is something to be avoided as stressful. Ignoring these kinds of journeys can provide a considerable barrier to public transport mobility access (Hine and Mitchell 2001). Research shows how moving with children engenders the struggle of battling with not just one but many other mobilities who are entirely dependent. Here the idea of the individual subject moving with others is supplemented further with various technologies and prosthesis added along the way. Christine Hine and Julian Mitchell's exploration of public transport exclusion highlights how negotiating public transport involves, for some people, the distressing management of a complex assemblage of mother, shopping, children, buggy which is made even more complicated with the task of counting one's change.

The trauma of mobility may even reside in the times when one is immobile 'waiting' for the bus to arrive. The inadequate provision of waiting spaces has been documented to produce further barriers to transport access. Back to Hutchinson's exploration of Los Angeles, urbanist Mike Davis' portrayal of the 'privatisation of public space' sees bus stops incredibly 'inhospitable to women riders, who are faced with the potential hazard of waiting for a bus at night' (2000: 114). In Britain and elsewhere similar experiences are to be found with 'waiting' spaces characterized by the traces of criminality and a lack of shelter.

The mobility impaired can face similar issues, especially for those who are reliant upon someone else to facilitate their mobility. The need and desire to move may mean someone is dependent upon a third party to help and augment their mobility. These third parties may in fact be technologies and a multitude of objects travelled with. Investigating the experiences of wheelchair users, Imrie demonstrates how the precise technology that liberates their mobility – the mobile prosthesis of the wheelchair – can bar their access to other places. Here we can quote from one respondent's description of a typical situation:

> the local bank have a ramp, but just try and get up it: no chance. So they've given me a service call transmitter I press it outside and a buzzer goes off inside and staff come and get me, so this is instead of having electronic doors and a decent ramp, what a waste.
>
> (Imrie 2000: 1652)

It may even be difficult for a third-party carer or helper to deal with the prosthesis and paraphernalia a person may need, from oxygen to wheelchairs and canes. The third-party friend, helper or carer may face restrictions on their autonomy. Carers may experience limitations and a sense of immobility as they are tethered to those they care for (Hanlon *et al.* 2007; Yantzi *et al.* 2007). Both carer and cared-for may well be further tied to the place-based nature of care (Angus *et al.* 2005; Dyck *et al.* 2005). Or else, for example, care givers of the opposite sex may be barred access to public toilets and washrooms and, therefore, their caring services are subsequently denied (Wiles 2003).

It appears that the hegemonic discourse of transport policy has been premised upon a 'universal, disembodied subject which is conceived of as neutered, that is, without sex, gender or any other attributed social or biological characteristics' (Imrie 2000: 1643). The paradoxical neoliberal ideology of freely mobile and productive bodies (Whitelegg 1997) can subsequently serve to construct immobile and marginalized bodies.

Development and displacement

The final example of mobile differentiation we can dwell on lies within an important area of development studies that takes the relation between development and mobility, or what has been located as 'displacement', very seriously. Jenny Robinson has written how there 'is increasingly a need to address how the dynamics of displacement also affect development' (Robinson and Mohan 2002: 2), encouraging what she calls a more 'mobile account' of development studies. Mohan and Robinson's recent book *Development and Displacement* forms a landmark volume that attempts to feel out the complex relation between social and economic development and mobilities and migrations (Chatty and Colchester 2002; Dutta 2007). While the authors recognize that 'development studies has few intellectual resources to analyse', they attempt to draw together 'a range of conditions of mobility – of people, and of resources and ideas' and through that, 'attempt to excavate their implications for development' (Mohan and Robinson 2002: 258).

One of the main issues scholarship has focused upon is how development agencies are able to respond to mass and forced displacement. We will consider later the impact of war on mobility, for mass migration and displacement may be caused by drought, disease and many other factors which create a vast need for food, shelter, resettlement and long-term repatriation. Development may stimulate and provoke mobility itself, what Robinson (2002: 3) labels the 'forced displacements undertaken in the name of development [. . .] development-induced displacements'.

There are countless historical precedents worth noting in the history of empire and national expansion. Movement across a space whose prior occupants are assumed to be dispersed and displaced is a defining characteristic of imperial exploration. In Australia *terra nullius*, the principle of no-one's land (Lindqvist 2007) made any space that was occupied by isolated individuals lacking in 'political society' available for discovery and occupation. Serving as the foundation for the massive displacement of aboriginal communities in Australia (Gelder and Jacobs 1998; Havemann 2005) and Canada, the legal principle instrumentalized the political authority and possession over land that was already occupied (Whatmore 2002), thereby legitimating its evacuation.

Today, large-scale development projects have had an enormous impact upon mobility and displacement, although smaller-scale projects such as mining, road construction and urban development (Turton 2002) have probably contributed more to overall displacement. Dams in particular are one of the biggest contributors of third world development to forced resettlement (Dreze *et al.* 1997). For Turton, borrowing from World Commission statistics, the overall displacement of world dams could reach something between 40 to 80 million people. In China alone 10.2 million people were forced to move out of their homes by dam building between 1950–1990. Such projects invoke enormous amounts of capital investment ranging between $32–46 billion in the 1990s. As a tool of development, dams contribute electricity, irrigation to nearby crops as well as supplying clean water for industrial use and urban centres.

Dams force people to move by appropriating land and property (Thukral 1992; Tharakan 2002). As the state exercises its rights of compulsory purchase orders, for Turton this sort of development can have 'disastrous consequences for their economic, physical, psychological and socio-cultural well-being' (2002: 51). They may include the loss of one's land and employment or housing. The displaced may suffer economic and social marginalization leading to under nourishment because they are unable to afford the most basic of foods. Other social ties and networks may be lost (Turton 2002: 50).

At the same time, this kind of displacement commonly occurs against groups and communities who are often the poorest and most marginalized members of the nation-state. Many development projects like dam construction occur in remote locations which are often inhabited by indigenous communities and diverse ethnic groups. As Fox argues, there 'is a direct association between large projects involving displacement and the lack of political representation of displaced peoples' (Fox in Turton 2002: 51).

Case study 3.4: DAMS AND DISPLACEMENTS

A clear example of the impact of dam construction can be found in the Kariba Dam in Zimbabwe, which caused the resettlement of 57,000 people between the beginning of its construction in 1956 to the completion of stage 1 of the project in 1958. In the planning of the resettlement of the Gwembi, Tonga people who lived in the local area were neither consulted nor involved in the decision-making process behind the construction of the dam or the resettlement of their own community. While there was an almost complete absence of political opposition to the dam by the Tonga, the resistance of one group of villagers saw 10 of their members killed by police forces. Furthermore, Turton explains that because the Tonga were not seen as 'stakeholders' in the dam project, little attempt was made to ensure that they would benefit from the dam's construction. The Tonga were only connected to the electricity national grid some 25 years later, and the best land for irrigation was given to an international company for development.

Further reading

(Dreze *et al.* 1997; Turton 2002; Dhagamwar *et al.* 2003)

MOBILE POLITICS

Let us start this final section of the chapter with two different styles of mobility. In the 2006 cinematic reinvention of Ian Fleming's James Bond in *Casino Royale*, the second scene of the film sees Daniel Craig chasing down a suspected terrorist across the uneven landscape of a building site. Bond's mobility is powerful. He busts through plasterboard walls and uses a digger to knock through a concrete one; he moves in fairly straight and predictable lines. Mollaka, the suspect he is chasing (played by Sébastien Foucan – the man with the contentious title of co-founding the art of parkour – or free running) moves with a rather different style. Foucan's terrorist runner doesn't dominate space, he doesn't break through it but he plays with it. His mobility finds new possibilities in the building being made. Mollaka leaps through windows, jumps up walls and hops between floors. Both Bond's and Mollaka's mobility work to subvert the ordinary norms of moving around the construction site. Yet they are still rather different. While Bond's movement takes apart the site and creates and dismantles space as he goes, Foucan's mobility works with what he has. Mollaka can't change the space, but he can find alternative ways of negotiating it.

The divisions I have made between both mobilities are rather simplistic and of course their utilization of space is far more complicated than I have portrayed; however, they do illustrate two far too common ways of describing mobile forms of conflict, contestation and violence. Bond appears to be the powerful dominator. He is chasing his suspect. He is destroying and unmaking space as he goes. Foucan's terrorist on the other hand is being chased. He is resisting arrest or worse by running. Unlike 007 his resistant mobility has to work with the space it is given.

Similarities can immediately be made with Michel de Certeau's (1984) idea of the 'tactic' which is rooted in such metaphors of mobility. Walking for de Certeau is an act of resistance through which spatial elements and formations may be questioned by local improvisation, just like Foucan's tactics of evading Bond in the film. On the other hand, the 'strategy' is a technique of the powerful. The strategy is enacted by those who shape and direct space just like Bond is able to do. As we witnessed in Chapter 2, these two kinds of mobility and positions of domination/resistance are common ways of understanding political struggle (Atkinson 1999). Mobility is often given meaning as a way to slip away from power and domination, and so we find that fluidity and the idealization of mobility has been taken up in a variety of places such as queer theory. Here 'great stock' is taken in 'movement, especially when it is movement against, beyond or away from rules and regulations, norms and conventions borders and limits' (Epps in Ahmed 2004: 152). But, it is not as simple as this. We will come back to our scene from *Casino Royale* in a moment, for we need to take in first what Cresswell has to say about the dualisms associated with mobile resistance and domination.

Cresswell (1999b) takes another film *Falling Down* to make his point. Describing Michael Douglas' character D-Fens' material negotiation of the LA freeway, the city and the private golf course, he rebukes any suggestion that mobility is a tactic only for the powerless. Cresswell's 'diagnostics' purports that mobilities always exert and evade power. Reading *Falling Down* like this means that D-Fens' ability to 'to walk and shoot his way through public space' may be read as a threatening masculine power or a defiance of the codified norms that overlay the golf course he crosses (1999b: 265–266)

In other words, mobility is not essentially resistance or domination; it is potentially both or either. Mobility is able to exert power that may well dominate, convert, contest and liberate (Cresswell 1999). For evidence of this we can look again at the relationship between mobile bodies in our chase scene. To say that Bond is the aggressor is far too simplistic. In a way, it is the terrorist who is the more dominant force. It is he who has set up this chase. It is he who evades and forces Bond into more and more dangerous positions, from lift shaft to leaping from cranes. And indeed it is he who initially wins by escaping to the embassy. In other words, both parties exert mobile power. Both resist and both dominate. Both performances of power cannot so easily be essentialized to their mobilities.

Throughout this chapter we have thought about some of the mechanisms by which the politics of mobility is enacted in terms of its relation to social difference and inequality. But this hasn't said a whole lot about the mobile composition of how one becomes political and exerts one's politics. What we have done instead is built up a picture of the physics of power that may push, pull, enable or constrain political relations. There is a danger that we see mobility strangled, sorted and denied, or overdetermined as an outcome of some other practice, regulation or policy. In this final section of the chapter we may look in more detail at more overt forms of what it means to *be* political by following the mobile composition of contestation through protest and violence.

Moving on

Consider first how mobility has very often taken centre stage as the contentious object of many protest movements. We could list a host of protests at various mobility infrastructures and developments from the construction of roads, toll roads and motorways, to the expansion and creation of airport terminals (Doherty 1998, 1999; Paterson 2000; Pascoe 2001). Mobility is in and of itself a controversial subject because of the effects it has. Enwrapped in a 'power-geometry' this is not really all that surprising when we begin to think about who may benefit from mobility and who may not. Mobility is very often a concern for the environment, and it is multi-scalar. From the environmental degradation caused by footpath erosion, to the noise of traffic and commotion in a nearby road, the destruction of a local habitat by the construction of a town's bypass road, to the contribution to global warming by an airport.

In addition to the object of opposition, mobilities compose political contestation. Mobility has always been an essential strategic tool in the practice of waging war. Without mobility, war could not take place. Mobility is necessary for the deployment of the army, for the long-range use of bombers and long-range missiles, for the swift movement of light tactical armies. The complex logistics of warfare has meant an address of the absolute 'position and juxtaposition' of things on their way through various chains of movement (Thrift 2004b). War has meant gaining a grip over multiple sequences of movement and traffic control of troops and recruits to bullets, fuel and food (see McNeill 1995 in the following chapter). Thus, mobility has taken centre stage in the strategizing behind the conduct of war. From Sun Tzu to Halford Mackinder, Liddell-Hart and the well-publicized doctrine of 'shock and awe' evident in the American bombing of Baghdad during the Iraq War, speed, surprise and flexibility are considered powerful 'kinetic' weapons (Virilio 2005).

War is made up of complex mobilities with critical implications for others. Caren Kaplan explains how war contains the contradiction of being 'one of the more perverse enactments of mobility in modernity' entailing the mobilities 'of large armies and instigates the mass displacement of refugees', while, at the same time 'it also polices borders and limits freedom of movement' (Kaplan 2006: 2396). Even as mobilities compose and constitute political violence they are often a vitally important component of its intended and unintended material effects. While there is little room to do any justice to this topic (such a focus would take up the space of many books), research can be drawn upon to discuss its violent and catastrophic effects which has led many people to seek shelter and safety elsewhere. Forced Migration Online (http://www.forcedmigration.org) have catalogued the numerous cases of mass displacement attributed to modern and recent instances of civil war and other kinds of state violence. Several Rwandan refugee crises occurring over the past two decades illustrate how vast political change and contestation have produced some of the largest mass migrations in recent history. 1990s Rwanda saw mass genocides and political unrest that led to the forced migration of more than 2 million people seeking asylum in Congo, Tanzania, Uganda and Burundi.

A product of war and violence, the mass movement of people and things has been recognized as a key strategy in the way warfare is conducted. In order to deliberately create disorientation, panic and especially mass hardship (Schivelbusch 2004; Bourke 2005), homelessness was a target during the Second World War bombing of Germany by the Allied Forces. The resultant firestorm in Hamburg left some 1 million German people homeless (Lowe 2007). In the case study box below we can see the evidence of how war necessarily involves the mobilization of meaningful materialities or what urban theorist Anique

Figure 3.5 The forced displacement of Rwandan refugees

Source: Copyright © Howard Davies/Corbis

Hommels (2005) would describe as the 'unbuilding' of the urban environment as well as human lives.

Case study 3.5: BULLDOZERS, BOMBS AND BRIDGES

The urban landscape is one of the greatest losers in modern warfare; notions such as 'urbicide' indicate how the city has taken centre stage as a target in war (Gregory 2004). Many writers argue that the city has become so important because of the recognition of the lives and relationships that go on in and through it. Croatian journalist Slavenka Drakulic's writings of the Bosnian–Croat destruction of the Stari Most, or Old Bridge in Mostar, Bosnia Herzegovina during 1993, urges one to think of the significance of such destruction. The bridge, which was 400 years old, had been essential to the life of the inhabitants of Mostar. It was a symbol of permanence, but it crucially facilitated daily mobilities, composing the rhythms and patterns of Mostar life. Destruction of the bridge did more than simply destroy a material thing. International Relations author Martin Coward suggests (2006: 420) that destroying the bridge ensured that the mobilities, lives and relationships it once supported became no longer possible.

Steve Graham's investigation of this practice set its sights on the long-standing Israeli–Palestinian conflict through which personal mobilities, lives and lifestyles are simultaneously disrupted as the urban structure is made mobile. The IDF utilization of the technology of the bulldozer has become common practice to wreak havoc on Palestinian settlements, by literally bulldozing away built structures and occasionally the occupants themselves. Of course Palestinian suicide bombers have similar capacities to 'unmake' Israeli lives on public buses and spaces. Detailing an interview of the Israeli Prime Minister Ariel Sharon – nicknamed the 'Bulldozer' by his compatriots – Graham quotes from the discussion as Sharon was asked how he would respond to Palestinian sniping at the new Jewish settlements of Gilo in the Palestinian neighbourhood south of Jerusalem, he answered:

> "I would eliminate the first row of houses in Beit Jela." The reporter enquired: what if the shooting persisted? Sharon replied: I would eliminate the second row of houses, and so on. I know the Arabs. They are not impressed by helicopters and missiles. For them there is nothing more important than their house. . . . It is better to level the entire village with bulldozers, row after row.
> (2004: 201)

Here Sharon emphasizes the attachment Palestinian occupations make to their homes. Demolishing the houses and moving the occupants away is far more effective for Sharon than orthodox military deployment. Indeed, the operational use of bulldozers permits the mobility of the army itself as bulldozers carve vast route-ways through the settlements so that following tanks can make their way through.

Further reading

(Graham 2002, 2003, 2004b; Coward 2004, 2006)

The walk

Mobilities have long been used as a means to both show up inequalities, power struggles and injustices or, elsewhere, to maintain them. The protest 'march' is a classic example of how people moving through streets, roads and cities works to subvert and contest power by the symbolism and significance of their mobility through space. Examples are wide ranging. Lucy G. Barber (2002) explores the evolution of Washington, D.C. as a theatre for political protest, witnessing numerous marches on the city to voice popular concerns. Mobilities in the city enacted spatial strategies that vocalized political issues and rejected state doctrines. Similarly, the decision to go to war in Iraq caused such public affront that hundreds of marches occurred in cities over the world on the 15th and 16th of February 2003, consisting of some 10 million people. But just as walkers form protest marches in order to take hold of space as a form of contestation, the parade and the pageant work with the apparently opposite aim of furthering state propaganda and dominance. Familiar demonstrations of military strength might see soldiers marching through streets and public squares in London or Moscow. In Singapore, the National Day parade has worked to appropriate, take hold of and 'invade' the spaces of everyday life by 'transforming ordinary streets into theatres of pomp' thereby engaging directly, the 'habitations of the people' (Kong and Yeoh 1997). Clearly the political significance of a march, parade or demonstration depends a great deal on context.

Marches do communicate the resistance or dissatisfaction with an established order and they can conversely seek to communicate and enforce it. But if we look closer and start to differentiate the march further, we uncover quite different kinds of mobilities of contestation. There is much more to protest and parades than symbolic display. Take Paul Routledge's (1997b) examination of protest in Nepal. In this example, protest movements enacted several different strategies of pedestrian mobility that varied in style, speed and number.

Case study 3.6: PACKS, SWARMS AND STYLES OF PROTEST IN KHATMANDHU

In this case study box we dwell on the work of geographer and activist Paul Routledge, studying the protest tactics held against the King in Kathmandu, Nepal in 1990. Resistance against the regime was enacted through a number of different mobile strategies. These techniques worked on two different horizons. They symbolically attacked the status quo through the significance of their movement as a visible confrontation to an established order. At the same time, their mobility actively and practically evaded capture by the authorities demonstrating how their mobility was much more important than the way it became socially significant.

Routledge explains that the key to this process was the way particular mobilities were able to occupy space. Protest tended to take the form of either a 'swarm' or 'pack'-like formations. The swarm, he argued, is large in number, 'effecting a movement of territorialization' (1997b: 76). These mobilities appropriated spaces and, therefore, power by their occupation of a meaningful territory. The movement of the swarm was, therefore, large scale and politically meaningful as an open and very visible contestation of power.

In contrast, another technique was adopted by the protestors that enacted quite a different form of movement in order to have quite different effects. What Routledge calls the pack was much less visible, and its contestation neither open nor direct. Instead, the pack would tend to be a lot more fluid, its use was sporadic and it would appear in a much more random regularity. Small groups of protestors would spring up in places in order to shout anti-government slogans, or they might burn effigies of the King. This would all happen incredibly quickly before the protestors would swiftly move on. As Routledge describes, the pack 'does not confront dominating power, it is more secretive, utilising underground tactics, surprise . . . their action always implies an imminent dispersion' (Routledge 1997b: 76). Protestors would even organize packs to form at the same time. They could then act as diversionary tactics to lead police away from more substantial gatherings.

Further reading

(Routledge 1994, 1997b)

Routledge's example illustrates how mobilities of protest go beyond mere symbolism. Mobility is not always enough to conjure powerful representations and messages. Rather, marches and movements must sometimes disrupt; they must alter normality in order to achieve their aim.

James C. Scott (1998) shows how the streets and spaces of insurrectionary politics can be used to support different sorts of mobilities in order to quash and disrupt rebellion. Scott explores Baron Von Haussmann's famous reorganization of Paris in the late nineteenth century. Constructing grand boulevards segwaying the city, Haussmann believed that the old narrow and windy streets of Paris enabled a subversive and insurrectionary politics of resistance. A nomadic population was supposed to inhabit these geographies with little connection to property. In order to oppose this unpredictable, invisible and mobile population, Haussman created a series of new avenues that linked the inner boulevards and the barracks on the outskirts of the city. The purpose behind this move was to enable free movement between different areas of the city. New roads could then enable more direct train and road transport between each district and the military outposts who would maintain order. As Scott shows, 'new boulevards in northeastern Paris allowed troops to rush from the Courbevoie barracks to the Bastille and then to subdue the turbulent Faubourg Saint-Antoine' (Scott 1998: 61).

Taking hold of space

What we are building up to is the idea that mobile contestation means more than the direct opposition or exertion of authority through visible and symbolic movements. Mobilities have often been figured as powerful forms of contestation by way of their embodied negotiation of urban spheres and environments. The activities of the Situationist International (SI) formed in 1957 and led by Guy Debord (1970) are key. Conceiving the urban as a sphere overrun by capitalist enterprise and visual mediation, the SI adopted a number of different techniques or tactics with which the spell of the urban spectacle could be broken (Macauley 2002). Debord's *psychogeography* sought to re-capture one's emotional and physical tie with the street – once lost through the distracting mediations of signs, symbols and imagery that both pacified and depoliticized. The situationists practised walking or 'drifting' through a *derive*. There are many parallels here with Certeau's formulation of walking noted above. Enacting a mobility of appropriation, by walking one could take hold of space and use it for their own purposes. This is an active production of space through its use – a 'spatial acting out of place' in order to manipulate and subvert ideological inscriptions.

Already seen at the start of this section in the movements of Bond's pursued terrorist, the practice of free running or parkour has become highly visible as an array of practices that disrupt the urban order (explored in more detail in the case study box below). The activity of

Figure 3.6 Parkour bodies

Source: Copyright © Saville

parkour is widely theorized by the people who practise it as a bodily resistance of urban architectures, unlocking space with utopian potential (Kraftl 2007; Saville 2008).

Case study 3.7: PARKOUR, UTOPIA AND THE PERFORMANCE OF RESISTANCE

Parkour or free running is the highly embodied forms of urban practice that revolve around inventive forms of mobility. Popular representations of parkour in James Bond, television documentaries and video games, highlight its emancipatory potential. Parkour is portrayed as a heroic disembedding of one's locatedness within the strictures of the contemporary city. Like the SI, there are easy comparisons to be made with De Certeau's account of the walker's appropriation of the urban fabric.

Although geographer Stephen Saville's (2008) fascinating exposition of the spaces of parkour suggests that we take these imaginings with some caution, he argues that the everyday practices of parkour are figured with utopian potential to transcend prescription. Saville explains how the philosophy of parkour is wrought with these notions. Even Sébastien Foucan (who played Bond's adversary discussed earlier) has written how parkour enacts an embodiment of imagining and dreaming. '[I]t's necessary to continue practicing, searching, travelling to discover, meet and share,' Foucan explains (cited in Saville 2008).

Yet for Saville, who undertakes his own participative ethnography of these practices, parkour is more than a finished job of finding freedom. His investigative (re)search for the process of parkour uncovers the activity as something akin to a search, or a questing, 'a search for new and more elaborate imaginings' (Saville 2008). Parkour appears to open out an opening; it is a cavity made 'out of possible, but not necessarily attainable, mobilities' and constitutes an unparalleled space by moving for play and creative engagements with cities and built architecture.

Further reading

(Borden 2001; Saville 2008)

Just like parkour, other forms of 'extreme sports' such as climbing have been described as a 'kind of corporeal subversive politics' (Lewis 2000: 65). Such mobile practices reject the everyday and the ordinary in search of extraordinary and heightened experiences. These examples figure practices that are formative of a utopianism. They appear to grant value to the potential collaboration of spaces and bodies that are generative of quite new ways of inhabiting and moving through city spaces.

Other sorts of mobile conflict have involved the construction of a political space in order to enable a similarly different future involving quite different sorts of movement and bodies. Airports, particularly, have been used for demonstrations against issues such as the building of a new runway. In December 2008, Bangkok's Suvarnabhumi airport in Thailand was encamped by several thousand protestors rallying against the government – shutting down all movements in and out of the airport for several days. To disrupt the mobilities and potential mobilities of a highway or motorway (see also Doherty 1999; Plows 2006), Routledge (1997a) shows how protest movements has territorialized space. Illegal mass trespass has occurred, such as the Pollock Free State movement which protested the new M77 motorway in Glasgow, Scotland, during 1995. Tactics common to direct action included the delay and the disruption of tree-felling activities by 'locking-on' to particular pieces of equipment immobilizing the road's construction.

Slow and fast

By now it is easy to see how mobilities have come to constitute political conflict, and particularly protest. Strategies of resistance and/or contestation often rub up against very similar strategies aimed at quelling or subduing these actions. As we have seen these seem to be premised upon not only movement but different styles of movement, mobilities of alternative direction, speed and predictability.

Mobile method 3.2: MULTI-METHODS AND THE SLOW FOOD MOVEMENT

As discussed in an earlier method box, walking is an important practice in the art of mobile ethnography. But while a mobile visual address of the city provides us with a useful method to engage in the flow of urban experience, it falls short in encompassing the multi-sensory experiences at the centre of mobile experiences and especially practices of resistance.

Sarah Pink's (2007; 2008) investigation of the Slow Food and Cittáslow (slow city) movement applied an approach more attuned to the movement's commitment to the sensual and slow appreciation of time and their belief in the importance of natural and environmental techniques of local food production. The movement has seen over 100 towns adopt its principles with networks established in Italy, the United Kingdom, Germany, Poland, Japan and New Zealand.

To allow her to experience the movement's activities Pink used an approach of an 'urban tour' that meant essentially walking the town and engaging with the physical fabric and the sensual qualities of its local goods.

Audio and visual methods were used to record Pink's journey as were detailed notes regarding the feeling, hub-bub and rhythm of the town. By taking part in walks and guided tours, drinking and eating while socializing in local cafes, Pink found that she could become attuned to the experience of the Cittáslow movement, savouring the tour as a slow 'wayfarer' (Pink 2007).

In practice

- The idea of an urban tour allows a direct and mobile engagement with a place

- Sharing common experiences of movement can enable a researcher to get closer to the rhythms and social experiences of that place
- Recording sounds, sights, smells and feelings can build a much fuller and dynamic reconstruction of mobility and the mobilities of a place

Further reading

(Pink 2007, 2008)

A useful instance of this dynamic is located in the petrol crisis I mentioned at the beginning of the chapter which disrupted many drivers during the summer of 2001 in the United Kingdom. The behaviour of panic buying had the unintended consequence of creating mass disruption at the fuel pump. In order to protest rising petrol prices and the shortages of fuel, various interested organizations such as road hauliers created their own go-slow convoys along various motorways and main roads through England, Scotland and Wales. As politics scholars Brian Doherty *et al.* (2002; 2003) examined, farmers and hauliers demonstrated at refineries in the Northwest, and in Pembrokeshire in Wales, while a slow convoy of 100 lorries and tractors created massive tailbacks on the M1. A day later, all the major refineries and oil depots had been blockaded in Wales and both the North and West of England: 'By Monday 11th September demonstrations at oil depots had spread to Scotland and the South of England' (2003: 4).

The go-slow movements and the blockades worked because of the impact of their mobilities and immobilities upon other people's ability to move. The disruption gained even more political freight as it was lifted off the context in which it happened; helicopter images of slowed-down and grid-locked main roads proliferated in newspapers and media imagery. Their mobilities and relative immobilities became mobile, mediated and circulated media vectors. The disruption spread. The police response to these movements took their own differently mobile course. The authors note how the police were able to manage and regulate the protestors by imposing conditions upon the convoy which had travelled from Berwick to London that 'greatly annoyed the protesters'. Police car mobilities, 'ensured that the convoy travelled faster than protesters intended and motorway exits were blocked in order to control the route' (2003: 15).

Similarly, Nick Blomley has shown how during the miners' strike in 1980s Britain the police developed mobile strategies of 'intercept and turn back' in order to deal with the techniques of protest employed by the miners (1994a, b). Blomley explains how a complex politics of mobility was constructed within the small spatial field of a striking picket line. In the context of the strike, picket lines formed outside the mines in order to produce a visible and public form of protest, as well as a physical site of disruption and disciplination to other miners. As we will explore in the case study box below, a nexus of ideologies, rights, and mobilities coincided.

Case study 3.8: MOBILIZING THE STRIKE

Nick Blomley's account of the miners' strike demonstrates the highly visible physicalization of many of the issues this chapter has dealt with.

The picket line itself acted as a 'muscular' and visible mobile embodiment of the conflict, as the line worked to constrain and deny movement and access to the pits. In crossing the line, a miner's mobility simultaneously constituted 'a disciplined action' that would conform to their own contract with their employer, while it would break trust and solidarity with the union and fellow workers.

These mobile dynamics were up-scaled away from the mine by the 'flying picket' that attempted to block, slow down or disrupt other workers making their way to other pits along various roads and motorways. As Blomley suggests, the mass flying picket was even more threatening than the relatively scaled-down and immobile picket line. It was '[p]remised on the alarming ability to move unpredictably, at will, en mass, beyond the disciplinary compass of the state' (1994a: 177).

How can we try to understand the police's reaction to the flying pickets? The police were able to directly oppose the flying picket with their own 'flexibility and mobility' (1994a: 160). They enacted their own differential politics of mobility by stopping the pickets and the union strikers in order to facilitate the mobility of the 'scabs' attempting to break those lines.

For Blomley, this configuration relied upon an entanglement of one's 'right to work' with one's 'right to move'. Without the component of mobility the 'right to work' had little value, composing only a weak complaint of the 'moral assuasion applied to strike-breakers by the union' (1994a: 175). As Blomley argues, 'the "right to work" was folded into another commanding liberty – that of free movement'. In the eyes of authority by blocking the scabs the pickets acted to constrain one's "right to move" (Blomley 1994a: 174).

Further reading

(Blomley 1994a; Cresswell 2001)

CONCLUSION

To move is to be political. Mobilities are underscored by political decision making and ideological meanings that arrange mobility and the possibility of mobility – motility – in particular ways to relations of society and power. Understood conceptually, mobility is placed within a complex geometry of influence as it shapes and forces and is shaped and forced by other actors and constraints. A politics of mobility thus demands an attentiveness to wider ideological assumptions about mobility as a social object and who or what should have access to it. It compels analysis of those who are cast in particular and unequal differential and hierarchical relations to mobility. As discussed, some may be in charge of mobility, while others are left behind or swept along by it.

The binaries of domination and resistance hold little water in this analysis as relations of power and control are enacted in much more plural and complex ways. Understanding mobility within the geometries of power it is formed through and comes to effect allows us to begin to visualize the often convoluted political relations it is involved in. From the simple access to services and the enabling of one's rights as citizen, to the complicated blurring of belonging by post-nationals, the waging of and protest against war, and the uncertain consequences of mobility for people a thousand miles away effected by rising sea levels, the politics of mobility is clearly multifaceted and incredibly contingent.

Importantly, the chapter ends with a case study illustrating the actions of miners and police during the miners' strike of 1980s Britain. Their movements gestured towards a natural division our discussions of mobility have run up against constantly so far. Both the miners crossing the line and forming the line seemed to constitute a mobility that had several kinds of signature. The (im)mobilities carried symbolic baggage. The enactment of crossing the picket line crossed something else. It crossed union solidarity and perhaps bonds, verbal promises or unspoken ties with friends and workmates. The mobility meant something. But it was more than that. This is radically simplifying the event, but it indicates what the following chapter will deal with head-on. Their mobility or immobility formed what Blomley referred to as a 'muscular' presence. In other words, the doing of these mobilities materially blocked, influenced or halted another. By going beyond the limitations of meaning or representation, in the following chapter we will ask: how is mobility done?

4

PRACTICES

Words expressing feelings, emotions, sentiments or certain mental and spiritual states will but touch the fringe of the inner responses which the shapes and rhythms of bodily action are capable of evoking.

(Laban 1960: 92)

[T]here is more in the action than meets the eye.

(Evans-Pritchard 1956: 231)

INTRODUCTION

In previous chapters we have concerned ourselves with ideas about mobility, be they from academics pondering what mobility means and how it works, or how mobilities are conceived out there in the world. While almost every kind of mobility *can* be given meaning and significance, from the turning of the globe to the movement of atoms, this does not say that much about the actuation of mobility itself: how mobility really happens. In this section of the book we can start to think about how mobility really takes place. To do so demands drawing upon a range of authors who have concerned themselves with the *doing* of mobility.

Let us be clear, however, that this does not mean subduing the importance of either the visual or the representational. Despite the fact that this kind of work has emerged as a definite critique of the primacy of the visual field, it is more about looking to what Lorimer (2005) calls the more-than-representational slivers of experience that draw attention towards the *doing* of the mobile – the enactment of the act and all that goes with it. Or for dance theorist Isadora Duncan, 'If I could say it, I would not have to dance it' (cited in Thrift 1997: 139). Thus, it is in the act of moving itself, in its enaction as an original and primary experience, which research on mobilities has sought to explore. The rest of the chapter is structured as follows. Initially we will dwell on three dominant but related positions or approaches from which mobilities have been examined as more-than-representational. Following that, the chapter then takes how these ways of thinking have been applied to mobilities at both more-than-visual and emotional registers. These examples explore the multi-sensorial and felt characteristics of mobility as they constitute many important social actions and phenomena. Let us first dwell upon why the *doing* of mobility is so often lost.

DOING MOBILITY

The anthropologist Brenda Farnell (1994) once wrote an article on representation aimed specifically at her own field of anthropology. Taking issue with the way various anthropological monographs and reports tended to represent the behaviour and activities of the subjects of their research, Farnell used Evans-Pritchard's classic monograph *Nuer Religion* (1956) as an example.

In the text Evans-Pritchard uses a photograph under which a caption reads 'Movement in the Wedding Dance'. Farnell argues that this image is instructive of the numerous ways that the moving body has been evinced from academic inquiry, namely, of the way movement or mobility, just as we saw in Chapter 1, has been frozen into snapshots. 'It is not uncommon,' she writes, 'to find actions reduced to a position or to a sequence of positions in this manner, such that a series of photographs, sketches, diagrams or positions of limbs plotted on a two dimensional graph are presented as records of movement' (Farnell 1994: 929). In mind of the search to look beyond this sort of snapshot explored in Chapter 1, Farnell argues that they do more than merely displace movement from its pictorial context. The removal of mobility by snapshots enacted an epistemological sleight of hand that 'removed the medium of bodily movement itself from serious consideration as a component of social action' (1994: 929). Mobility was subsequently less than a serious subject for the advancement of academic research, nor was knowledge of it adequate enough to stand as research findings, data or information (Farnell 1996, 1999). More seriously, she argued that the removal of mobility in images either reflected or worked to further compromise 'anthropological inquiry by distorting our understanding of ways of knowing and being' (1994: 929) that revolved around mobility. '[M]any socio-cultural anthropologists (although they are certainly not alone in this) literally do not "see" movement empirically' and as Farnell goes on, 'when they do, it is conceived of as "behaviour" rather than "action"' (1994: 936). In other words, the abstraction of mobility from its contextual meanings serves to relegate it to an involuntary and behavioural reaction, layered beneath or beyond the remit of society and culture.

Farnell went on to situate this failing within the wider anthropological concern to uncover systems of meaning performed and communicated through the medium of the body (1999). '[T]he body, albeit a social and cultural one rather than a biological or mechanistic entity,' she wrote, 'remains a static object' (1994: 930). Farnell wanted to know much more about 'accounts of persons *enacting* the body, that is, using physical actions in the agentive production of meaning, actions that may be either out of awareness through habit, or highly deliberate choreographies' (1994: 931). Farnell sought to uncover the 'production of meaning' through the moving, mobile body as we have seen in Chapter 2, but more importantly, she wanted to recover the enaction of mobility – by the moving body as 'physical actor' – as a core component of the social world.

Farnell's arguments (Farnell 1994, 1996, 1999) are particularly useful in articulating two key points this chapter will deal with. These concern how mobilities as meaningful social actions may occur through unthought habitudinal movements, and, moreover, how a fuller and more embodied sense of mobility may evade abstraction as well as photographic and even textual representations. What I am trying to get at here is how various writers have argued that mobilities are something more, something more than what we can read from the pages of a book, our interpretation of a wedding dance, or what one may tell us about it.

Let us take a rather simplistic yet useful example. If you have played golf before try to remember what it is like to take a shot. If you don't play golf, you might think about another sport. Consider dancing or, say, kicking a football. If you can't do that, just think about wiggling your little finger up and down. If I were to ask you to describe your body movement, how easy would that be? You might be able to give me a fairly convincing description of the movement of your technique: how you moved. But this is quite a superficial take on movement that really only captures a thin slice of the experience. Is how your body moved from a to b, from one position to another, really all there is to it? Nigel Thrift uses Wittgenstein's famous formulation in order to pose the question: 'what remains over from the fact that I raise my arm when I subtract the fact that my arm goes up?' (Thrift 2000a). Have a go and what are you left with? What did moving feel like? What sensations did it throw up? How did it *make* you feel?

I have just had a go at this myself, the bits I could do with confidence were the parts that I can see and imagine. The basic and actual mechanical mobility of my arm is not a taxing thing to describe; it is something I can visualize on paper, something I can almost separate from reality and draw with lines and a few arrows. Remember Cresswell's (2006a) diagram of movement from A to B? If you have had a go at this experiment I warrant that trying to describe the other aspects of moving, the feelings and sensations you felt, might be a much harder challenge. Be aware of the fact that you are rarely *aware* of your mobilities either. From fidgeting to driving yourself to work and forgetting how you got there, to smacking a tennis ball in anger, you are never really cognizant of all the movements you may make. Yi-Fu Tuan (1975) uses the example of driving in order to question the proposition that we construct so-called mental maps that allow us to guide and be guided along our daily life-courses, finding our way home, to work or wherever. For Tuan, all of this can happen quite unconsciously as he finds himself at a considerable distance from where he first started with no recollection of how he got there, his mind being seemingly elsewhere during the journey.

It is from this kind of position that a theoretical turn has emerged in the social sciences which orbits around the terms performance, practice and the concept of 'non-representational theory'. These can help us understand mobility in rather different ways to those of Chapter 2. While the example of my arm does a rather poor job of explaining the issue, researchers in this area want to examine just what is left over from the mechanical representation of an arm moving, a body mobile. As we have seen there is a universe of experiences, feelings and sensations to be got at.

PRACTICE, PERFORMANCE AND MORE-THAN-REPRESENTATIONAL MOBILITIES

Scholarship has attempted to move beyond the primacy of representation and meaning towards both body-centred experiences and forms of knowledge that are rooted in philosophical traditions such as phenomenology. In this section we will examine three interrelated approaches that have sought to understand mobility in ways that complicate and question the relation between mobility, styles of thinking and ways of representing.

Habits and practices

The phenomenology of philosopher Maurice Merleau-Ponty (1962) has become an incredibly popular approach to recover the indivisibility of mind and body. Ponty's investigations of various psychiatric disorders sought to understand the body in action, which, he thought 'should enable us to arrive at a better understanding of it' (Merleau-Ponty 1962: 117). Ponty rejected the Cartesian division between mind and body in favour of what he described as body subjects. These are subjects who experience the world in a way which is phenomenal – an address of the world that happens before any reflective and conscious thought can occur. Rejecting the assumption that consciousness or representational thought determined intentionality, in this active and pre-cognitive understanding, the mobile body does not sit within a simple container of space and time, but it actively assumes them as the body becomes a direct intermediator between subject and world.

In geography and elsewhere, explorations of movement from behavioural and psychological fields have posited quite different explanations of mobility that look at what factors provoke or cause it to take place. In the following key idea box we focus on David Seamon's interjection in this debate as he questioned cognitive and behaviouralist approaches to mobility in the late 1970s.

Key idea 4.1: PHENOMENOLOGY AND THE MOBILE 'BODY SUBJECT'

Humanist geographer David Seamon's (1979) contributions stand out as important interventions into the role of bodily practices and mobilities in everyday life, battling against approaches drawn from behaviouralism and cognitive science.

Inspired by his contemporaries and especially the writings of philosopher Maurice Merleau-Ponty, Seamon argued that scholarship needed to take greater notice of not only human experience, but human experience as it took place. He took the assumption that experiences were apprehended directly through the body as his starting point.

Taking up Ponty's notion of the body-subject, Seamon investigated the centrality of habitual and mundane bodily mobilities in the formation of everyday social relations and social places (see below). It was through these studies and his interpretation of phenomenology that he rebutted many other theories of mobility at that time. According to Seamon (1979) behaviouralists were concerned with the body's movements, but saw the mobility of an action like driving as 'a collection of reactions to external stimuli'.

On the other hand, a cognitive approach (Downs and Stea 1974) would argue that 'this apparent "stimulus–response" sequence is not so simple'. Downs and Stea suggest how 'in these situations you are *thinking ahead* (in both literal and metaphorical sense) and using your cognitive map' (Downs and Stea cited in Seamon 1980: 153). Seamon's answer to these problems was to turn to Ponty's conception of the body subject. This approach envisaged how the body could transform regular needs and behaviours into habits that 'meet the requirements of everyday living'. Overcoming the cognitive presumption that everyday mobilities would require mental maps or reflective thought, paying continuous attention to each 'gesture of the hand, each step of the foot, each start' (1980: 156) was not necessary. The body subject sees the mobile body as more than a set of nervous reactions as a behaviouralist might, but rather 'an intelligent, holistic process which *directs*' and thus, overcomes any notion of the body as 'a collection of passive responses that can only *react*' (1980: 156).

Further reading

(Downs and Stea 1977; Seamon 1979, 1980; Buttimer and Seamon 1980)

Seamon looked into the *practice* of mobility – how mobilities occur in ways that compose meaningful social actions. His ideas were clearly influenced by the now-familiar humanistic concern for the stability of significant human existence (as seen in Chapter 2). Seamon went on to explore how certain 'preflective' bodily movements act as a bodily 'stratum of life'. Our lived spaces, everyday routines and habits appear to be made up of micro-gestures such as 'stepping, turning, reaching' that may add up to daily activities, or agglomerate into meaningful routines and significant places and environments.

Washing dishes, ploughing, house building, potting, hunting, boiling the kettle, putting the coffee pot on, are the combination of small-scale body movements that have *fused* together into a recognizable and repeatable practice that achieves a particular task, end or need. These practices are composed of simple 'arm, leg, and trunk movement' that become aligned or even 'attuned' to the completion of tasks and work, and importantly, they appear to 'direct themselves spontaneously' (Seamon 1980: 158). By studying these mobilities, one should gain 'a picture of the stabilizing habitual forces of a particular lifeworld' (1980: 162).

From this style of approach, these body movements were blended into even larger behaviours that Seamon described as space–time ballets. The idea that the sense of stability or 'continuity' of micro-movements that are repeated spontaneously is upscaled even further to compose the patterning of daily street life. It is worth quoting from Jane Jacobs' (1962) classic *Death and Life of American Great Cities* which Seamon refers to.

> The stretch of Hudson Street where I live is each day the scene of an intricate sidewalk ballet. I make my own first entrance into it a little after eight when I put out the garbage can, surely a prosaic occupation, but I enjoy my part, my little clang, as the droves of junior high school students walk by the centre of the stage dropping candy wrappers ... While I sweep up the wrappers I watch the other rituals of the morning: Mr Halpert unhooking the laundry's handcart from its mooring to a cellar door, Joe Cornacchia's son-in-law stacking out the empty crates from the delicatessen, the barber bringing out his sidewalk folding chair.
>
> (Jacobs 1962: 52–53)

In addition to the sorts of social street scenes that Jane Jacobs evokes, other theorists of mobility argue that mobilities enable the reproduction of social orders. In this approach, mobilities do more than conjure up meaningful social encounters through habitual and unconscious actions, but they work to repeat and reinforce social ideas, norms or ideologies *because* they are taken as habitual and unthought. It is here that we may turn to our second key idea for the chapter so far, from the late French sociologist Pierre Bourdieu.

Key idea 4.2: MOBILITY AND THE *HABITUS*

Bourdieu's remarkable studies of tribal societies in Algeria, *Outline of the Theory of Practice* (1977) and *Distinction* (1984) have become celebrated texts in the analysis of culture. Bourdieu suggested that societal norms and values could be internalized and repeated through the body's movements, practices and routines – by what he called the *habitus*. These he described as 'principles, practices and representations' which could be regulated and adapted to goals without the need of conscious aiming 'and collectively orchestrated without being the product of conscious direction' (1977: 77).

Focusing on bodily mobility, Bourdieu provides us with some interesting examples of male and female dispositions towards movement which he suggests demonstrate bodily *hexis*. For Bourdieu this was 'political mythology realized, *em-bodied*'. The norms and values of the local societies he looked at were 'turned into a permanent disposition', as the repetition of the body's mobility formed durabilities of particular ways of feeling, thinking and moving. These norms are especially gendered.

Bourdieu showed how the biggest differences in gendered mobilities were displaced through attitudinal dispositions to movement. For the female *centripetal* disposition, their movement was very inwardly directed, leading towards the house and the hearth. For men, their mobility was alternatively *centrifugal* leading outwards to the market and to the fields. A man 'knows where he is going and knows he will arrive on time, whatever the obstacles, expresses strength and resolution' (1977: 94).

On the other hand, the woman's centripetal attitude meant that she might 'walk with a slight stoop, looking down, keeping her eyes on the spot where she will next put her foot . . . her gait must avoid the excessive swing of the hips' (Bourdieu 1977: 94).

Importantly, these dispositions were seen to be beyond the 'grasp of consciousness,' thereby making deliberate transformation almost impossible (1977: 94).

Further reading

(Bourdieu 1977, 1984; Thrift 1983; Cresswell 2002)

The 'centripetal' description of female mobility is repeated elsewhere, albeit from diverging philosophical perspectives. Questioning the universalism of Ponty's phenomenological approach more closely than Bourdieu's, Iris Marion Young's (1990) famous essay 'Throwing Like a Girl' comes to the similar conclusion that feminine bodily movement appears centripetal, passive and 'self-referred'. Denying Ponty's exploration of the body and its motion as an undifferentiated and primordial intentional act, like Bourdieu, Young finds how women refrain from mobilizing their whole body into motion but rather concentrate movement into 'one part of the body alone' (1990: 148). By moving only one part of the body, the relatively immobile part anchors and even drags the rest of the body down. Young argues that feminine motion is frequently contradictory, being made up of mobile parts moving against frustratingly fixed ones – leaving their movement circuitous and wasteful. Female motion is much more inward facing. The woman is supposedly the object of motion as opposed to its originator and, therefore, she is disposed to act as a consequence of that motion. These actions are uncertain as she does not feel her body's 'motions are entirely under her control' (1990: 150) and must divide her attention between the task at hand and inducing her body into action. Finally, feminine movement is performed in order to be observed; it is a thing that is 'looked at and acted upon' (1990: 150).

From 'running like a girl, climbing like a girl, swinging like a girl, hitting like a girl' (1990: 146) Bourdieu and Young see how differently gendered bodies address the world quite differently through their mobility. Whilst for Young this differentiation appears below the registering of the social, Bourdieu pitches its exertion in much more of an overt way. Mobilities are reproductive of a series of social norms, values and ideas about being a woman or a man. Such ideas are repeated in popular myths and cultural assumptions as shown in Mary Gordon's study of the interaction of female and male mobility in literature, wherein, 'The woman is the centripetal force, pulling the hero not only from natural happiness but from heroism as well' (Gordon 1991: 15). Here we have a coeval intertwining of mobility and society. Yet all three approaches mark out mobility to be what they see as primordial, pre-cognitive and certainly unintentional acts.

Performance and non-representational theory

We have seen how relatively stabilized forms of mobility both add up to, reflect and reproduce an apparent social order. Habitualized mobilities generated relative permanencies of meaningful places and the sedimentation of repetitive encounters in Seamon's humanism. Alternatively, Bourdieu accounted for bodily mobilities that reflected and reinforced the political mythology of *hexis*, reproducing relations such as gender divisions. Drawing upon similar styles of thought, other approaches have taken different ends of the equation. Attending to the *moment* of action – an attention to the act of *doing* – and by emphasizing the same unconscious and pre-reflective posturing of intentionality found above, they question how mobilities exceed our capacities to even think about and represent them.

The great majority of this sort of writing has been developed in fields such as dance and performance theory exemplified in Rudolph Laban's (1960) articulation of the non-representational characteristics of the mobile body performing an activity such as dance. Laban's interest in the inner-worldly feelings of dance led him to compare its articulation to poetry translated into prose (1960: 91). Finding both processes entirely unsatisfying, Laban considers the movements of chopping wood, embracing or even threatening someone. These movements, he argues (1960: 92), have little to do with the symbolism of movement, 'Man in those silent movements, pregnant with emotion, may perform strange movements which appear meaningless, or at any rate inexplicable.' Any attempt to portray these movements through words might only touch the 'fringe' of the experience, as 'Movement can say more, for all its shortness, than pages of verbal description' (1960: 92).

There are important themes in Laban's words that speak to the intangible *something* in the act – the *doing* of bodily mobility that surpasses our capacity to explain what that is through pages of verbal description. Laban catalogues a host of feelings, emotions, states and experiences that are formed through mobility but are very difficult to articulate. This counts not only for the subject doing the performance of moving, but for the spectator watching the event of the performance of the act.

In this sense the performance of mobility may be 'nonreproductive', it cannot be re-presented because something is lost from the original. The performance of mobility is suggestive of a now-ness to the act of moving that performance theorist Peggy Phelan describes in the art of spectatorship. For Phelan, there are no 'leftovers' as the spectator struggles to absorb all that they can see, hear and feel in 'a manically charged present' (Phelan 1993: 148). There are elements to the performances that Phelan describes which, like the movements I asked you to think about earlier, escape representational forms of capture or knowledge. Representing the performance by a photo, or say a video tape, records only a partial print of the innumerable and complex dimensions of the performance itself such that it 'ceases to become performance art' (1993).

You might try having a go at this yourself. Video yourself while mobile – kick a ball, jump as high as you can, sprint down a hill, or simply raise your hand. Replay these motions. What do you get? You might remember from the images and sounds something about the experience and something no doubt about the place or the landscape in which your activities took place. But what is missing? Is the feeling of the wind in your face represented in the images? Is the kinaesthetic sensation of moving as fast as you can portrayed? Pain may be etched on your face as a grimace, but does it recreate the sensation of the burning of your muscles or the exhilaration of speed? Probably not. As Loïc Wacquant describes in the context of boxing, 'How to account anthropologically for a practice that is so intensely corporeal, a culture that is thoroughly kinetic, a universe in which the most essential is transmitted, acquired, and deployed beneath language and consciousness' (Wacquant 2004: xi). Dance, running, skipping, jumping, a multitude of mobilities enacted as part of cultural practices, sports, games, gestures and role-play, all of the movements produce experiences that have, 'no meaning outside of a world of sensations, of movement, of the loss and recovery of physical control' (Radley 1995: 4). They are, in other words, a 'physically sensed way of being' (Thrift 1997: 148).

Attending to the momentary event of mobility in action does more than criticize attempts to abstract and represent it. What J. D. Dewsbury has described as a 'flash of the felt before' it becomes wrapped 'up in contemplation' (Dewsbury 2003: 1930) seeks to ask the difficult question of how mobilities might be thought. How might they be intended or directed by conscious or unconscious cognition? We know from Seamon and by way of Jane Jacobs' writings on street life that no matter how unconscious they may be, habitudinal ballets of mobility can have important and meaningful consequences.

Take one's capacity to learn how to control his or her own bodily mobilities. Some of us know about the kind of struggle at play between mind and body when attempting to accomplish a difficult or complicated bodily motion. Sociologist Loïc Wacquant shows how in the context of boxing: 'It is the trained body that is the *spontaneous strategist*; it knows, understands, judges and reacts all at once' (2004: 97). Training one's body as an accomplished boxer is to, therefore, learn to be mobile without thinking. The comparison of a trained and learned body with an amateur, means that novices are easily recognizable by their 'stiffness and academicism' which 'betray the intervention of conscious thought into the coordination of gestures and movements' (Wacquant 2004: 97). What is learned, in other words, may be only 'rendered visible in the act of doing' (Dewsbury 2000: 472). Sam Keen's (1999) extraordinary reflective volume on learning to be a trapeze artist highlights the weaknesses of representational knowledge. Keen accounts for how he tried to find his way, 'out of my head and into my body – the path of sensation' by developing what he calls his kinaesthetic intelligence. Keen explains how he had to 'abandon concept, analysis, image and word' but to feel directly into the signals and impulses coming from 'muscles and nerve endings'. Creating an 'immediate, intuitive awareness of where my body was in space. . . .' Keen gradually felt a way into his 'body-in-motion' (Keen 1999: 147). Learning to be mobile in this way was not necessarily accomplished by seeing, by reading, by consciously taking on knowledge, or purely thought about through words, diagrams and ideas. Keen's abilities as a trapeze artist were rendered through an intuitive awareness generated by doing and experience of this doing.

Wacquant's ethnography of boxing demonstrates how the representational directions of mobility are explicitly rejected. This tension could be found in a moment when he discussed the use of a boxing manual to his coach and trainer DeeDee. Here I quote from Wacquant's fascinating account:

> As I'm drying myself with a towel I let slip: "Hey DeeDee, you know what I found in the library on campus the other day? A book called *The Complete Workout of the Boxer*, which shows all the basic movements and exercises of boxing. Is It worth reading it to learn the fundamentals?"
>
> DeeDee screws up his face in disgust: "*You don' learn to box from books. You learn to box in d'gym.*"
>
> "But it could help you to see the different punches and to understand them better, no?"

> "No, it ain't helpful. You don't learn how to box readin' no book. I know them books, they got a buncha pictures an' diagrams in'em that show you how to place your feet an' your arms, the angle your arm is supposed to move at an' all that, but it's all standin' still! You don't get no sense of *movement*. Boxin's movement, it's the movement that count." I persist: "So you can't learn anything about boxing in books then?"
>
> "No, you cain't."
>
> "But, why not?"
>
> In a tone irritated by my insistence, as if all of this went so much without saying that it was useless for him to repeat himself: "*You just can't! Period.* You can't. In a book everything's standin' still. They don' show you what's happenin' in d'ring. Tha's not boxin' all that stuff, Louie. You can't, tha's all."
>
> (Wacquant 2004: 100–101).

In lots of instances, the right moves, the right techniques of mobility might be read, understood and learnt from manuals. No doubt this occurs in boxing all of the time, yet in this instance the representations were rejected. It was the embodied, kinaesthetically motional and original – in the moment – quality of mobility which DeeDee wanted Loïc to experience. By 'standin' still' – without experience – he could never learn the right technique of being mobile. This example is not meant to suggest that the practice of movement was simply unthought, just that simple conscious thought directed towards that particular activity was not necessary for the learning experience.

Mobile method 4.1: THE REPRESENTATION OF RESEARCH

The question of how we represent our research engenders some strong consideration in relation to the notion of mobility. Geographer Sarah Whatmore (2003) borrows from science and technology theorist Isabel Stengers to question our conceptions of the pristine 'field' that we go out to. A common perspective in research methodology is that the field or field site of one's interview, ethnography, observation or social survey is a passive world – *out there* – that lies down and plays dead for researchers to capture and probe, as if the world was stopped (Whitehead 1979; Ingold 2000).

Such an issue may be traced forwards from the anticipation and event of data collection through to the representational strategies of research writing. In anthropology the idea of an inert or pristine world has been dominant as shown in Mary-Louise Pratt's (1986) critique of the 'arrival trope' persistent in many anthropological monographs. The arrival into a tribal camp is consistently romanticized and fictionalized in academic discourse that renders scenes of 'sunrise', 'mounds of blankets and animal skins' with the awakened villagers 'stoking the coals, rebuilding the fire, and warming themselves in the chilly morning air' (Shostak in Pratt 1986: 43).

The scene presents 'an ethnographic utopia'; it is 'a traditional society doing its traditional thing, oblivious to the alien observing presence' (Pratt 1986: 43). This is an unspoiled and clearly divisible world that cannot be. Conceiving the field as a fluidic, active, complicated and messy terrain that is encountered in surprising and uncertain ways would be a more accurate portrayal that is sympathetic to the mobilizing process of the research itself. 'Energetic exchanges' are set in motion by research practice (Latour 1999). Practices which literally put the world into play (or perhaps more at play, see Ingold 2000) see researchers becoming 'interinvolved' by taking evidence back to the lab, compressing earth and soil underfoot, bringing gifts and new ideas to welcoming strangers, or simply causing pedestrians to cross over the road to bypass their questionnaire!

In practice

- Be aware of the mobilities your research has set in motion
  - What have you brought to the encounter?
  - What have you taken away?
- During the analysis and write-up of research it is important not to romanticize and clean up what was probably a contingent process

Further reading

(Pratt 1986; Thrift 1999; Ingold 2000; Whatmore 2003)

More-than-representational mobilities

Several authors writing on these issues have questioned whether non-representational theory has over extended its attention to a pre-cognitive domain of action and becoming. And in fact, they have suggested that non-representational theory has worked to foster the kinds of dualisms that it was originally meant to move beyond. Scholarship has questioned whether it draws an implicit line between thought and action and between the social and the 'unanalysable world of the precognitive or prereflective' (Nash 2000: 657). Catherine Nash goes on to argue that it is only by pretending that dance is above any sort of social space, 'by imagining dance as a free-floating realm of the experiential' below or above any idea of a social and cultural sphere, that it can 'be thought of as a prelinguistic and presocial bodily experience' (Wolff in Nash 2000: 658).

Debates surrounding mobility have intervened in this discussion in several ways. Some scholarship has focused on the romantic proposition that bodily non-representational mobilities somehow elude the hand of power. Cresswell's work discussed in the case study box below is probably the most notable example to question this kind of assumption.

Case study 4.1: REPRESENTING AND REGULATING MOBILITY IN MODERNITY

On the Move provides one of the strongest criticisms of non-representational theory. Cresswell sets about discovering various moments in Western modernity when mobilities have been abstracted and represented. Of particular note are the systematic processes to make mobilities knowable before that knowledge is re-inscribed back upon the mobile body through various forms of bodily investigation and rationalization.

From the Victor Sylvester dance school in London to the emergence of scientific management of the body in Taylor's steel factory in Bethlehem, Pennsylvania, and even the application of Lillian Gilbreth's analyses of the home, are just some of the moments Cresswell gives us. Attempts are consistently made to make mobility knowable, to capture mobility by various imaging devices that made the invisible visible. Time–motion studies placed incomprehensible body movements with a temporal signature (Rabinbach 1990), allowing the break-down of smooth motions into a mechanistic assemblage of component parts.

Once the many different ways of abstracting mobilities had been completed, the knowledges of mobility are subjected to various judgements and calculations that correspond to and reinforce various social values, principles and ideals. Made commensurable with the turnover time of capitalism, bodily mobility could be cast as inefficient, slow, lazy and even animalistic. Placed on a par with racial dispositions and associations, Sylvester's dance school laboured to outlaw improper dance steps. In other words, social meaning was never far away as it inscribed motions and steps, which we have seen marked with connotations of free expression, with the taint of the proper and improper, or right and wrong.

Further reading

(Crary 1999; Bahnisch 2000; Solnit 2003; Cresswell 2006a)

Cresswell contends that mobility has almost always been governed by representations, and is, therefore, unable to slip away from power. Various systems of capturing and representing mobility, from movement notation to time–motion studies displace bodily experience into an abstractive quantity to be measured and calculated. Treating mobility in this way allowed the body to be understood, rationalized and inscribed with meaning. Certain mobilities were accepted while others were not. Mobilities were learnt and governed by representations. Bodily movement such as dance was not a straightforward denial of relations that sought to restrict pure play and pleasure but instead became 'part of the play of representational power' (Cresswell 2006a: 74). To describe dance as a mobile bodily action that is only and purely 'non-representational' risks separating out relations between thought and action, between representation and presentation. Cresswell's answer to move beyond such divisions is to suggest that 'human mobility is simultaneously representational,' taking 'practical representation as practice and practice as representation' (Cresswell 2006: 73).

I think we should be very sympathetic to these perspectives while being wary that this kind of criticism may overplay the divisions writers of non-representational theory have appeared to create. Derek McCormack's (McCormack 2002, 2003, 2004) studies on rhythm are actually quite careful to avoid the kinds of dichotomies charged against them. Drawing on Gregory Bateson's writings on dance, McCormack's investigation of rhythm questions the tendency to separate out thought from mobile action by asking what it is that we mean by thought itself. Quoting Bateson,

> Isadora Duncan, when she said, "If I could say it, I would not have to dance it," was talking nonsense, because her dance was about combinations of saying and moving.
>
> (McCormack 2002: 439–440)

It is not that the bodily mobile practices McCormack examines are above or beyond thinking or representational thought at all – lying purely in a domain of precognition – but that they require or involve different sorts of understandings of thinking and feeling that are implicated with one another.

This is an approach which is not limited to representational thinking and feeling, but a different sort of thinking-feeling altogether. It is a recognition that mobilities such as dance involve various combinations of thought, action, feeling and articulation. Tim Ingold proposes a similar rethinking of walking when he suggests that 'cognition should not be *set off* from locomotion, along the lines of a division between head and heels' (2004: 331). Walking, in this sense, is part and parcel of unconscious bodily percepts and feelings while it is itself a kind of 'circumambulatory knowing' (Ingold 2004). Consider further how walking forms part of other incredibly thoughtful practices of scientific measurement, data collection, survey or mapping as Lorimer and Lund (2004) show in their study of highland mountaineering. Walking is both thought and unthought.

Mobility appears to be both simultaneously representational and non-representational. Mobilities are often given meaning and they are very often represented. They can be very well thought through, training and bodily fitness being an obvious example of thought and mobility (Latham 2008). Representations may even be used to track, trace and control mobilities. At the same time, however, there are elements of mobility that are almost impossible to represent, that escape meaning, that occur without thought, but not necessarily power. Further still, just because mobility is difficult to represent, does not mean that people have not tried to. What is important is that we do not forget the coupling and co-existence of these two facets of mobility.

In the next two sections of the chapter, we will now take a sample of several dimensions and examples the *practice* and *performance* of mobility. We will place particular emphasis upon the sensations and feelings of mobility that, as we have seen, are tricky to articulate and represent.

SEEING, SENSING, MOVING: EXAMPLES OF MOBILE PRACTICE

In this first section we will deal with how mobility as *practice* and *performance* has been bound up with a host of different sensory experiences which may constitute, compose or for some even begin to *explain* mobilities. The first to be dealt with may feel the most familiar from our discussions of the *flâneur* who visually absorbs the sights and spectacles from the position of their wanderings. Mobility is, therefore, first understood as a visual activity.

Seeing

Perhaps one of the most in-depth accounts of mobility as a visual practice has come from the sociologist Ervin Goffman (1961) who undertook a fantastically original observation of walking in urban environments. For Goffman, the mobilities of the urban realm were directed and led by the eyes. Goffman's scopic focus assumed how mobile walkers engaged in the act of scanning; walking was, therefore, 'almost exclusively' taken as a '*visual* activity'. Goffman's pedestrian is pre-figured as a pilot – 'supposed to use his eyes to guide his body about' (Ingold 2004: 327). Like Goffman, psychologist James Gibson's (1950, 1979) work on ecological perception proposed the theory of visual affordances, where the visual perception of a given situation was supposed to 'afford' or draw out people's movements through a given space.

Gibson argued how an open space or an 'environment affords locomotion in any direction over the ground, whereas a cluttered environment affords locomotion only at *openings*' (1979: 31). When the visual field was blocked by obstacles, mobility was less likely. In attempting to explain this behaviour psychologically Gibson suggested that walking was a form of 'visually controlled locomotion' where '[a] *path* affords pedestrian locomotion from one place to another' (Gibson 1979: 36). Described more fully under his theory of 'natural vision and movement', Gibson's approach highlights a pre-cognitive visual way to apprehend the environment around us. Natural vision structured one's propensity to move if it appeared that there was the space to allow it. When no constraints were put on a visual system, 'we look around, walk up to something interesting and move around it so as to see it from all sides, and go from one vista to another. That is natural vision' (Gibson 1979: 1).

Gibson's approach of natural vision has found purchase beyond psychology through efforts to model and predict human movement in urban environments by planners, architects and software such as *Space Syntax*. Fields like pedestrian dynamics attempt to model or simulate the behaviour of mobile individuals. In questioning the characteristically a-sensual appreciation of mobile subjects in urban modelling, Turner, Penn and others have turned to Gibson's 'natural movement' theory as a way to incorporate visual perception into their simulations (Hillier *et al.* 1993). Questioning both the accuracy and dominance of modelling simulations that treat moving pedestrians like atomistic particles as considered in Chapter 3, the authors ask, 'does the corporeal human bump through a crowd of corporeal humans, or does the human guide him or herself through gaps in the crowd?' (Turner and Penn 2002: 474). What troubles Turner and Penn is that existing models lack the 'ability to see. Without it, their models can only be, literally, stabs in the dark at able-sighted human behaviour' (Turner and Penn 2002: 474).

By directly encoding Gibson's theory of visual affordance into their 'exosomatic visual architecture', the agents (mobile individuals) of the simulation are now given the capacity to see. An available walking surface affords a mobile pedestrian the opportunity to occupy that space, nearby walkable surfaces will draw mobilities on. Their hypothesis states:

> When engaging in natural movement, a human will simply guide him or herself by moving towards further available walkable surface. The existence of walkable surface will be determined via the most easily accessed sense, typically his or her visual field.
>
> (Turner and Penn 2002: 480)

Elsewhere, Hillier and Hanson (1984) propose that someone negotiating an urban space will be influenced by their 'line of site'. The longest line of site or 'axial line' will determine their most likely path of movement.

Discussions such as these are important. While they are intentionally simplistic, they demonstrate some of the rather instrumental purposes to which our ideas and imaginations about the practice of mobility are put – just as we saw in the previous chapter. These imaginings of mobility illustrate a relatively common conception that mobilities occur somewhat unconsciously and predictably according to one's perception of the world by their visual senses. We can turn towards others who have explored how the relationship between mobility and visuality is never purely an optical endeavour and neither a simple predictable practice. Moreover, many diverse cultural and economic practices that we might presume to be particularly visual, are shown to be made up of various forms of mobile multi-sensorial activities. We can consider first the art of the tourist.

Site-seeing

The mobile practices of the tourist have been located as a key visual activity that encompasses particular ways of seeing places, locations and people, all while being on the move (explored in more detail in the box below). As we saw in Chapter 2, the centrality of the visual in tourism has been appropriately captured by film and media theorists who posit the film and televisual experience as similar to that of the tourist's. Explored in more depth in the following chapter, 'travelling theory' conceives 'an image of both TV programs and TV viewers as travellers, tourists, sojourners, exiles, vagabonds, pilgrims, or nomads' (Harrington and Bielby 2005). From their couches, sofas or cinema seats, spectators may be transported to other times and spaces (Friedberg 1993; Bruno 2002) – the Taj Mahal, New York, Iceland, Baghdad – which the viewer hungrily consumes.

We could explore how tourism is primarily an activity of the visual senses (see box below), just as the film and television spectator could be taken as a subject transfixed by his or her ocular perceptions. Yet while the notion of the 'tourist gaze' looks as if it implies a singularly visual activity, both tourism and televisual cultural studies have demonstrated how site-seeing is constructive of a kaleidoscope of other sensory mobile experiences.

Key idea 4.3: THE TOURIST GAZE

Seeking to examine tourism as 'one of the defining characteristics of being "modern"' Urry highlights how being on holiday involves a series of movements and pauses: staying and going, leaving and arriving, which both come to constitute tourism as a status symbol of modern societies. Normalized as the thing one does with their summer vacation, these mobilities have a distinctively visual flavour. For Urry, there are a host of visual mobile practices through which tourism is composed and around which it is organized. He calls this tourist gaze.

Even before travel commences, a tourist may be subjected to images and visions of his or her destination, stimulated in anticipation through 'daydreaming and fantasy' (1990: 3) and passed on through other mobile media such as tourist brochures and television broadcasts.

The tourist's gaze is further an art of capturing and collecting. Images of places travelled, snapshots of experience, allow a reverse process of re-presentation. The holiday may be recounted to friends, or remembered at other times, or revisited during a return trip. The tourist gaze is furthermore a look on the extraordinary and out of the everyday. Tourists may seek difference, surprise and the unexpected. And yet these surprises may be premeditated or predetermined. The spaces of tourism are often very scripted, and the tourist gaze is directed and led towards certain things by signs, symbols and other imagery. For instance, the end of successive wars in the Balkans and the former Yugoslavia has enabled the repackaging of Eastern Europe's tourism. The wars now provide an object for history and heritage, while their cessation sells rural escapism as well the cosmopolitan experience of Prague and Sarajevo.

The final issue for Urry is that all of the activities add up to 'one of the characteristics of the "modern" experience' that is intensely mobile and visual (1990: 4). As we saw in Chapter 2, the tourist consumption of images comes to symbolize the mass commodification and consumption of just about everything.

Further reading

(Urry 1990; Sheller and Urry 2004; Hannam 2008)

Whilst Urry's approach does not preclude other kinds of mobile experience, the tourist 'gaze' has come under criticism for its prioritization of visual modes of mobility. The 'gaze' and the way it is constructed by various practices of seeing works to secure tourism mobilities as a field apart from other kinds of embodied leisure. Critics contend that it is the whole of the body and not just the eyes that see, experience and compose tourist activity. Indeed, Urry's later writings reflect these dimensions. Tourist scholar Pau Obrador Pons puts this point incredibly simply; he argues that it is, 'in fact, the entire body that is involved in tourist dwelling' (2003: 57). One could imagine a host of embodiments associated with the tourist experience from smells and touches to excitations, thrills and fears.

Pons explains how tourist practices are 'deeply grounded' in other sorts of mobile activities. The classic tourist setting of the beach is a case in point; for example, a spectacle of skin and sand (Fiske 1989; Shields 1991) or even the red and orange horizon of a sunset – the psychedelic rave continuing until sunrise – is a typical beach scene of Goa, Thailand, or Ibiza, where, much more than a visual space, the beach becomes 'a sonic tourist environment'. It is visual tourist activity 'constituted through the material connections to bodies, space times and objects' that a particular sort of music enables (Obrador 2003: 57–58).

This kind of approach moves us closer to the more-than-representational styles of mobility discussed earlier. In considering the whole body's enrapture in the spaces and practices of the mobile tourist we can consider the 'active, expressive and sensual' experience of a body in motion. This does more than explore the social and representational significance of values, norms and ideals of what, for instance, this mobile body might mean, but considers its fleshy, tactile 'experiments and desires' where a beach rave becomes a 'place of embodied utopias and nondiscursive pleasures' (Orbrador 2003: 55). In a classical music concert we might expect to 'see musicians bodily producing the music' as the performance is received by the audience who listen 'thoughtfully, silent and still'. Yet in the very different venue of the Goan or Ibizan beach, this relation between audience, performer and mobility is quite different; 'physical movement' becomes 'a necessary part of what it means to listen' (Frith in Malbon 1999: 84).

Urry clearly saw how many other sensations were paramount to the tourist experience. The significance of the gaze was that it prioritized the visual and, perhaps more importantly, it established and prescribed a series of practices and dispositions in order to consume these experiences. Tourist practices and doings are often made up of a bank of unreflexive dispositions as well as other practiced and embodied norms about how to act in a tourist stage. For Tim Edensor, these act as 'guides' to 'performative orientations' and 'a working consensus about what to do' (Edensor 2001: 71). They might even be formalized in textual form such as guidebooks which tell us what to see, what to touch, where to stand, which museum to go to and which restaurant to eat at. Let us take Edensor's example of a British tourist wandering around the site of the Taj Mahal. Edensor writes how she had become agitated by the behaviour of nearby domestic tourists:

> With exasperation she exclaimed: 'I think Indians are really crap tourists. They just don't know how to be tourists, rushing around, talking all the time and never stopping to look at anything – even here at the Taj Mahal!'
>
> (Edensor 2001: 78)

In this context, the tourist habits of the British tourist were suddenly exposed because of the differences apparent in her Indian counterparts. The exposure to Otherness suddenly made her own un-reflexive practices of being a tourist visible as a norm from which to compare others (Frykman 1996). For the British tourist, acting correctly meant the unquestioned performance of visual appreciation, collection and contemplation. Tourist spectatorship is clearly not a universal.

Thinking more critically about how one *should* see has led other scholars of 'site-seeing' to re-imagine the visual consumer in a far more mobile and embodied fashion, whose vision or gaze is structured and called to attention (Crary 1999). Giuliana Bruno notes how previous accounts of spectatorship have tended to overstate the immobile visuality of spectator experience as 'a motionless subject, enraptured in a state of solitary reverie' (Bruno 1992: 114). Similarly, Crang is very critical of the idea of the spectator-tourist rendered like 'living couch potatoes . . . bombarded by thousands of images crossing the living room each day' (Crang 2002b: 14). Erwin Panofsky, the German art historian, explained how the spectator-subject 'is in permanent motion as his eye identifies itself with the lens of the camera which permanently shifts in direction and distance' (Panofsky in Friedberg 1993: 126). Furthermore, Bruno's (2002) later work examines cinema-going as a thoroughly multi-sensorial experience happening with other spectators (Jancovich *et al.* 2003).

Mobile method 4.2: TIME–SPACE COLLAGE

In view of our discussion concerning the limits of representation and the problems associated in representing the experience of mobility, we are left with the question of how mobile methodologies can account for this excess. Indeed, if mobile social practices are conceived to be 'noncognitive, and in large part nonverbal, how can they be included within research?' (Latham 2003: 2001). How, in short, (Latham 2003) do we get beyond the matter of *talk*?

Striving to take the 'flow of practice' seriously, Latham puts forward a way to rework relatively traditional forms of research method from time geography. Attempting to convey the sense of fluidity and feel of his respondents' mobilities through the city of Auckland, Latham combines quotes from diaries and diary interviews his respondents recorded, alongside diarists' photography. Constructing a collage of quotes, photos and commentary from the research, the diagrams are placed on an axis of temporality and spatiality. The x-axis denotes the spaces of home, work and the Ponsonby Road, while the y-axis contains the scale of clock time. The connective tissue of arrows tie the activities of the day together and, indeed, form a major part of the experience of the day, leading 'the reader into the diarists' world'.

In practice

- Consider how the focus on *talk* belies unthought and habitual motions of activity
- Ask your respondents to present their own time–space diaries of movement
- Diaries, informal photos and notes can provide effective ways of representing everyday mobilities

Further reading

(Latham 2003)

Walking

As we have seen so far, attention to the other sensory components of seeing hints at arguments taken forward forcefully by geographer John Wylie as we return once more to the mobile practice of the walk. Inspired by the phenomenological writing of Merleau-Ponty, Alphonso Lingis and Tim Ingold, Wylie (2005, 2007) has provided a significant exploration and critique of the mobile gaze. Following Ponty's rejection of the Cartesian separation between mind and body, Wylie (2002) witnesses an 'enlacement' between the mobile walker observing the landscape and, indeed, the landscape itself. From this point of view, we can begin to reject, as Ingold does vociferously, the idea of a walker who merely skims 'the surface of a world that has been previously mapped out and constructed' to an approach that sees both landscape and walker as constructive of each other (Ingold 2004: 328–329).

Paul Adams (2001) writes how walking through a place demands an 'involvement' with senses such as 'sight', 'hearing', 'touch', 'smell', 'the kinetic sense called proprioception' and even 'taste'. Sounds could 'range from the calls of birds to the sounds of traffic and horns' whereas the sensation of touch might 'include the brush of tall grass, the spray of passing cars on wet roads, and the jostling of strangers in crowded places' (2001: 188). Wider philosophical writings have even argued that mobility and movement are composites of a manner of tactility. The capacity of touch to communicate feeling is animated by mobility itself, so argues architectural theorist Dalibor Vesely (2004). He explains, for example, how we

> pass our fingers over the table-top and apprehend its smoothness as a quality of the object. The tactile impression results from the completion of the movement. When the tactile movement stops, the tactile impression dies out.
>
> (2004: 82)

The negotiation of public space by the visually impaired highlights the importance of the tactility of the walk through the landscape (Hetherington 2003). The touches of a walk in a place like a museum can augment visual or non-visual wayfinding behaviour.

Importantly, these senses may not even be consciously thought through. Kevin Hetherington (2000a) discusses how visually impaired wayfinders may navigate space without thinking about it. The involvement of mobile touch can take place in the prereflective domain unthought 'like the beat of our heart' (Vesely 2004: 82). Various senses of touch and tactility, as discussed by Lewis (2000), can thus supplement our other senses. Representational knowledge such as a map may be combined with what Lewis calls tactile navigation, wherein, 'the climbing body locates itself by *feeling* its way through the world' (2000: 76). In this way, 'for the climber, knowledge is *grasp*-ed' (Lewis 2000: 76–77).

It is perhaps a cliché to say that one is changed along the way of these touchy mobilities. Common folklore tells us how people come down from a mountain changed. Spending a night at Cadair Idris near the Welsh seaside town of Aberystwyth is meant to turn one either mad or into a poet! Nonetheless, the idea of mobility as a translation finds its expression in John Wylie's account of the 'involvement' between himself and Glastonbury Tor. Taking a personal journey of smells, tastes, touches and subjectivity, the box below explores his ascension.

Case study 4.2: ASCENDING GLASTONBURY TOR

Wylie's ascent of Glastonbury Tor near Bristol in the United Kingdom combines personal narrative with conceptual analysis in order to question the standard visual and discursive interpretation of landscape. Wylie shows how climbing up the Tor is not simply a matter of seeing, nor seeing *from* the Tor.

Accounting for the various sensations and feelings encountered along the way, Wylie shows how moving up the Tor is an act of enfolding. His lofty visual perspective is formed through the 'heaviness one feels in chest and legs' (2002: 451); at other times, it is combined with a 'growing lightness, a sense of anchorage being slipped'. The gazing mobile body merges with the landscape it sees from and *with*. The enactment of a 'folding' envelops the subject within the landscape 'like the occlusion of two weather fronts' (2002: 451). From this occlusion both mountain and subject emerge as translations – as travelling companions who are altered along their journey. The climber 'emerges as a viewer' and the more durable mountain, slightly different.

The act of seeing and moving are entirely caught up in one another, just as they are with the mountain. It is too simplistic to say that one is simply *changed* by the movement, but the idea is there.

Further reading

(Ingold 2000; Wylie 2002; Ingold 2004; Wylie 2005)

For Wylie, the walk is a not necessarily a conscious process constantly being reflected upon, but seems to involve 'an ongoing milieu of folding and unfolding, intertwining and diverging' of sights, smells, touches, slips and falls, and it is through all of this that 'subjectivity and meaning' may emerge (2002: 445).

Cycling

The visual is supplemented as one dimension among many other senses in the mobility of cycling. Suggesting that the preoccupation with the visual is a parallel preoccupation with representation, Justin Spinney's recent exploration seeks to understand the non-visual and non-representational dimensions of the cycling experience. Spinney's relationship with his bike sees the technology as more than a tool to simply move the body through space. '[W]e feel from it,' Spinney writes (2006: 729).

Spinney's wonderfully evocative descriptions allow us to imagine the experiences of incredible heat, pain, rhythm and sheer will. Quoting from his participation in the Mont Ventoux cycle race, Spinney recounts the experience from his ethnographic diary:

> Breathing, staying in the saddle, getting out of the saddle, using different muscles, starting to hurt, but not unbearable, just going into the shade . . . calves starting to hurt, trying to find a rhythm, entering the shade, breathing very laboured, breathe, 9 mph, into second gear, one gear left . . . out of the saddle, corner coming up, looks like a ramp, all I can see is road . . . thighs burning.
>
> (transcription, 12 June 2003, Spinney 2006: 724)

Spinney's chain of sensations helps to communicate the visceral bodily actions of mobility. His story is enlivened by an account of the performance of the ride that was simple, intuitive and pre-reflective. We learn how his body reacts without thinking to the patternation of shade, alterations of the road's incline and changes in the tarmac's contours.

Pause for a moment and consider one of the strongest sensations articulated here: pain. As we will explore in the next section, the sensations and experience of mobility are often intertwined with feelings and emotions. In the context of running, John Bale (2004) asks, 'Can the pain endured by athletes be a perverse source of pleasure?' (91). The answer, for Bale, is that it is often constructed as an inevitable, and, for some people, an objective of running and the purpose of mobility itself. Pain can be part and parcel of accomplishing a personal best, or winning a race. In this sense, pain is a means to feelings of pleasure. For others, it is the other way round where pain and mobility become enwrapped in a form of masochism. '[T]he more it hurts, the harder we try to run' Bale observes (Bale 2004: 99). While pain might be part and parcel of Spinney's cyclist experience, his actions are not necessarily thought out or deliberated upon. Aspects of the body's mobility emerge through non-cognitive action and without deliberate intention, or visual recognition. The total bodily muscular consciousness, for Spinney, signifies a 'reprioritisation of the senses where the visual is relegated in the project of movement' (2006: 724). Clearly, some parts of his journey are incredibly visual and thought through, while others are not.

Bringing these three examples of site-seeing, walking and cycling back together, we can see how a preoccupation with the more than visual invites us to engage with all sorts of percepts and sensuous knowledges occurring with the brain, at times, seemingly switched off. That said, these sensory tales of mobility do not always mean cognitive disengagement. The activation of the body's muscular consciousness can enable more intense forms of thought and calculation, just as mobility brings to life a sense of touch. Philosopher Jean Jacques Rousseau conceived of the art of strolling as an act of thought itself (Van den Abbeele 1992). The mobile practice allows his 'soul to be released . . . and his thoughts to become more "bold"' (Van den Abbeele 1992: 114). Rousseau put forward that walking held something special that could, like a light switch, help turn on, 'animate' and 'enliven' his ideas. He wrote, 'I almost cannot think when I stay in place; my body needs to be in motion for my mind to be there' (1992: 114). As opposed to the corporeal stability required for Descartes' meditative journeys, Rousseau's locomotion of the mind could only be triggered by the locomotion of his body. Thought and motion were so intertwined that, 'as soon as I stop, I stop thinking and my head goes only with my feet' (in Van Den Abbeele 1992: 114).

Mobile method 4.3: TALKING WHILE WALKING-WITH

The 'self-retrieving' of Rousseau's walk has intriguing implications for qualitative research methods concerned with the recovery of memories and associations. Drawing on the peripatetic tradition of body movement and pre-figuring emerging notions of walking-with (Pink 2007), Jon Anderson (2004) explores the potential for 'talking while walking' as a suitable technique of research to examine environmental direct action (EDA). Excavating his respondents' memories, values and associations for the wild flower meadow they were assembled to protect at Ashton Court in Bristol, Anderson worked on the premise that by walking with a participant certain configurations of mind, body and world could come into alliance. Quoting from Rebecca Solnit's (2000) *Wanderlust*, walking

> is a state in which the mind, the body, and the world are aligned, as though they were three characters finally in conversation together, three notes suddenly making a chord.
>
> (2000: 5)

In Anderson's case, walking took the form of the mobility of the 'bimble', or what Evans describes as, 'to wander around aimlessly. Like "amble" but sounds more twee' (Evans in Anderson 2004: 257). Such movements generated a sympathy between Anderson, his respondent, and the environment they both moved through. The bimble enabled a collaborative encounter. Moving with the respondent in an environment that had considerable significance enabled the provocation of memories previously unstated or unrecalled while other inspirations, excitations and emotions could be prompted. As Lee and Ingold go on to document, by walking with, 'we can see and feel what is really a learning process of being together, in adjusting one's body and one's speech to the rhythms of others, and of sharing (or at least coming to see) a point of view' (Lee and Ingold 2006: 83).

In practice

- Engaging your respondents in a mobile activity such as a walk can provoke memories and sensations otherwise untapped
- Consult your respondent and carefully think about the potential effect of particular social and physical settings upon your encounter

Further reading

(Anderson 2004; Bassett 2004; Lee and Ingold 2006; Hein *et al.* 2008)

MOTION AND EMOTION: AFFECT AND THE FEELING OF MOBILITY

Just as we can take mobility as something we feel according to our various capacities to perceive and sense, mobility is something which may be *moved* and something we might be *moved by*. In other words, mobility is something we feel in an emotional and affective sense. Some philosophers and social theorists have attempted to illustrate the intimate connection between mobility, emotions and affects. Giuliana Bruno finds a causal interactivity between physical motion and emotion, whereby 'motion, indeed, produces emotion' and 'correlatively, emotion contains a movement' (Bruno 2002: 6). Similarly, for cultural theorist Brian Massumi affects and mobilities are inseparable; the mobile body 'feels as it moves and moves as it feels' (Massumi 2002: 2). The smallest movement of the body, 'convokes a qualitative difference . . . it beckons a feeling' (2002: 1–2), whereas feelings may demand correlative movements. Elsewhere, sociologist Mimi Sheller (2004a) has argued that, 'Motion and emotion . . . are kinaesthetically intertwined and produced together through a conjunction of bodies, technologies and cultural practices' (Sheller 2004a: 227; see also Ahmed 2004). Moods and emotions have even been taken as movements in and of themselves. Gaston Bachelard describes how our hopes and fears have a 'vertical differential' in how they might make 'us lighter or heavier' (Bachelard 1988: 10). Positive emotions may involve an 'awareness within ourselves of a release, a gaiety, a lightness' (1988: 10). Whereas more negative emotions of fear and anxiety involve a 'journey downward' or a 'fall'.

Although there are subtle differences between conceptions of emotion and affect which we will come on to later, movements of the body seem to be able to summon up feelings, which in turn may interfere with, aggravate, supplement or supplant one another. Likewise, it appears that emotions and moods can stimulate mobilities. Perspectives such as these have drawn on the theories of emotion famously developed by William James (Robinson 1998). In his renowned example of a man encountering a bear, James asks what happens when the man runs away. Is the emotion what causes us to run – we are afraid, therefore, we run? On the other hand, are we afraid because we run? You might ask the similar question, 'Do I cry because I am sad, or do I feel sad because I am crying?' There is not such a simple answer. A complicated chain of processes of emotion that are entirely bound up in the body's movement appear to be at work. James argues that the feeling is not the cognitive experience of an emotion, but the feeling is the running away. From this perspective, the feeling of being frightened is bound up in the mobility of running and in the qualitative emotion itself. One would not feel that feeling of being frightened without the mobile action of running away.

No doubt you can associate a host of feelings and emotions with all sorts of mobilities. Sheller takes the act of driving as a particularly powerful form of auto-mobility that can 'impress' upon specific bodies in different ways, and thus produce differing 'impressions' (Ahmed, 2004) or differing affective dispositions. The movement of driving and its associated senses of the moving view, the feeling of the breeze, the transition of movement, the thrum of the engine, the forces of turning round a sharp corner at speed, all these can produce a variety of feelings from 'happiness, excitement or anticipation; others become fearful, anxious or sick to the stomach' (Sheller 2004a: 227).

Many forms of mobility are intended as a way to experience and generate particular sorts of feelings. Le Breton (2000, 2004) describes the different sorts of adventure sports as a 'search for stress'. Feelings are not merely singular either, but certain kinds of mobility may disturb chains of emotion that follow one after another or feed back into each other. Before the 'jump' of a bungee jump, the 'first step' may translate into a 'feeling of giddiness' to panic, which may push 'emotion to the limit' and express itself as 'a shout' (Le Breton 2000: 4). Alternatively, mobile feelings may be a little more consistent, achieved by more sustained forms of movement. Many of the kinds of moods, feelings and states discussed so far are difficult to represent. Runner's high has been described as an 'unreflective, lived, culturally specific, bodily reaction' and, therefore, 'cannot be explained by accurate representation' (Bale 2004: 105). Quoting from Roger Bannister's autobiographical account of his experiences, Bale shows how one can only get a very marginal sense of what Bannister must have undergone.

> Every once in a while, when I'm running, I feel a sense of tremendous well-being come over me. Everything about me feels in harmony. I feel smooth, and my breathing is so relaxed that I get the feeling I can run forever. I'm not aware of time or space – only a remarkable sense of calm.
>
> (Bannister in Bale 2004: 106)

'Becoming smooth' is a feeling that perhaps we can only really understand or get a sense of by experiencing it ourselves.

As Thrift (2000b) suggests, understanding these feeling-states is important because they are often enrolled into wider economies and industries selling sensations, feelings and experiences – from adrenaline-filled adventure tourism, to theme parks, to cars. In the rest of the section we will trace two different ways of apprehending mobilities, affects and emotions. While they have emerged from rather different sets of philosophical and disciplinary contexts, we will see how they render imaginations of mobile bodies that are incredibly open to transmission, communication and participation.

Affects, affections and capacities

Our first treatment envisages the same sort of micro-physical universe of atoms bustling about that Lucretius introduced us to at the beginning of the book. Following his decomposition of everything to bits and pieces of atomized materials, Lucretius describes how our moods appear to emerge from the subject's correspondence with its environment. Our feelings, our 'vital spirit' appear to be constituted by the smallest of molecules that make us up, and, therefore, they are the 'first to be stirred'. Many of us know how susceptible we are to the changes in something as simple and enveloping as the weather (Ingold 2005, 2007a). Lucretius argued how our emotions are 'caught up by warmth and the unseen energy of wind, then by air' (1951: 103) and 'everything is roused to movement.' Receptivity is key; feeling subjects move as if a candle fluttering against a faint breeze. Upon stimulation, 'the blood is quickened' and like a line of dominos falling over, 'the impulse spreads throughout the flesh; last of all, bones and marrow are thrilled with pleasure or the opposite excitement' (1951: 103).

The susceptible bodies that Lucretius envisages point to both issues of representation as well as free will. Movement is either generated by the 'heart', mobilized 'throughout every member of his body', or it may come from without, perhaps prompted by 'a blow inflicted with compulsive force by someone else'. In this case, all 'the matter of our body is set going and pushed along involuntarily' until checks and balances may be imposed setting the body still (Lucretius 1951: 68). Like Lucretius, Spinoza alludes to a body always under assault from waves of objects and other bodies and things.

Motion and emotion. Cause and effect. One after another, the Spinozan imagination sees bodies affecting other bodies endlessly on and on (Gatens 1996; Damasio 2000; Anderson 2006). We have bodies bumping around. Bodies circulating and colliding. They are building up or diminishing, exciting or subduing, and in their collaborations they form capacities – capacities to move and capacities to feel. Take a recent exploration of these ideas by Sara Ahmed. Ahmed explores the capacities of a feeling such as fear. Fear, as we have already considered by way of the bear, is generative of capacities to be mobile. Fear may also diminish and shrink what it is that the body may do. For Ahmed (2004) in fact, fear involves a double capacity, for as '*it restricts the body's mobility*' it also '*seems to prepare the body for flight*' (2004, 69). There is a complex politics here, a power geometry which seals the fate of one's fear with the mobility of another. Ahmed draws on Fanon's story of a White boy running away from a Black man who passes by. The boy runs into the arms of his mother as Ahmed explains:

> we can see that the white child's apparent fear does not lead to his refusal to inhabit the world, but to his embrace of the world through the apparently safe enclosure formed by the loved one who fears the white child's fear, who is crushed by that fear, by being sealed into a body that tightens up, and takes up less space. In other words, *fear works to restrict some bodies through the movement or expansion of others.*
>
> (Ahmed 2004: 69)

In this example, the mother composes with her child's body. The fear of the child forms their rush to their mother's embrace, which they join with and now move through the world with. Yet the mother's empathy for their child's emotion means that they take on that fear which is 'crushing', immobilizing and restrictive.

Ahmed's presentation of the racial motivation of fear and mobility (White flight would be another example) can be compared with racialized assumptions about these capacities. Bale (1996) explains how Black athletes were widely perceived as 'born sprinters' as a result of their apparent trained and primordial instinctive capacities to react and 'run away' from danger. Quoting from a typical explanation of the African's competence for instinctive mobility, 'His speed "off the mark" and the maintenance of rapid action for short distances are, *without doubt* due to the primitive reaction to jungle instinct' (cited in Bale 1996: 143). Other dispositions are absorbed into a substrate of ethnicity and nationality. Bale goes on to quote from a German writing in the 1930s who, in describing the emotional temperament of the Finnish, illuminates the author's racial politics. Apparently holding the propensity to run deep within their blood, when a Finn is exposed to the 'clear, deep' and green forests, or the 'wide open luxuriant plains' or the 'heights covered by massive clusters of trees' one is overpowered, overcome and enticed by inner feelings of 'elation', a 'deep need', and 'you want to run' (1996: 143).

Communication and community

How do these emotions and affects become mobile in themselves: how does affect move? Much of the work to deal with this question has focused in on the mobile performances of the body as well as its integrity. Following the work of thinkers such as Emily Martin (1998), the body has been figured not as a 'fortress' that is closed off to outside forces but whose skin is analogous to a permeable membrane. According to Teresa Brennan's (2003) neuro-biological examination of affect, this means 'lifting off the burden of the ego's belief that it is self-contained in terms of the affects it experiences' (2003: 95). In this final area of discussion, we will explore how emotions and affects rise and surge between bodies. Bodies extend out into more-than-personal bonds and associations as people move with each other.

Emotions and affects feed back as they leap between people tying them even closer together.

The example of the Goan beach party mentioned earlier provides us with a useful way in. You may take part in ritualistic and regular mobilities yourselves – going out and dancing in a club. You may have been part of a parade, the armed forces or a variant; I am imagining drills and marches. You might have once or regularly taken part in a Mexican wave at a sports ground or cheered with your fellow supporters at a goal in football or a point in basketball. Can you think of what these movements felt like? Some of you might recollect feelings of solidarity, a feeling of the group – of something bigger than you. Some of you may have felt a feeling of pride, a kind of increase in morale or a respect for those nearby. You may have felt an innate connection to those around you, a sense of belonging. Or perhaps it was a sort of high, a feeling of exhilaration that rises up between yourself and those nearby (see box below).

Case study 4.3: MOVING TOGETHER IN TIME

Historian William McNeill's (1995) book *Keeping Together in Time* argues that people moving together in time has provided the affective glue to bond societies and cultures collectively. We will look at the performance of dancing in more detail in Chapter 5, but in this instance McNeill compares the energetic exchanges of dancing with other forms of rhythmical movement such as military drills. Exploring how dance has often worked to stabilize small and isolated communities in small tribes and villages, McNeill argues that 'the emotional arousal of dance (and less energetic forms of rhythmic movement like stately processionals and military drill) was fundamental in widening and differentiating social bonds among our species' (1995: 65). Rhythmic movement helped to create senses of belonging within communities and townships as well as military situations.

Moving in rhythmic unity to create a 'primitive and very powerful social bond' has become a popular technique to arouse troop solidarity. Drilling requires large groups of soldiers to carry out movements in perfect time and rhythm in order to benefit not just group coordination and synchronicity but a more intangible reward. Moving in this way created 'an intense fellow feeling' or a 'primitive reserve of sociality' often named as *esprit de corps* or morale. For many military commanders, the maintenance of morale was incredibly important to ensure that troops would respond to instructions, turn away from desertion, and that they would be 'ferocious' in battle.

From this sort of military drilling, to the tribal dancing of island communities, to the moves, shapes, bumps and grinds, you might witness every Friday night, McNeill sees that they all employ essentially the same principle.

Tracing these practices over the past 1000 years, McNeill concludes it is exactly these sorts of bodily movements which hold people together. They are so important that complex human societies could not maintain themselves without this 'kinaesthetic under girding'. Rejecting the power of ideology and discourse, it is feelings for McNeill that really matter, feelings which 'are inseparable from their gestural and muscular expression' (1995: 152).

Further reading

(McNeill 1995; De Landa 1997; Gagen 2006)

Rather than communicated symbolically or discursively, being mobile together in time is 'critical in both establishing and enhancing a sense of collective purpose and a common understanding' producing feelings of 'well being' (Brennan 2003: 70) in sports events and other communal gatherings. Understood without separation from the physical and practical movements of the body, feelings may extend beyond and envelop a group moving in time. Moving in accordance brings about senses and feelings of solidarity and belonging without verbal, communicative and symbolic forms of action.

These kinds of bonds and fellow feelings do not have to happen by moving in time, but by simply moving with. We have already seen how moving together has meant precisely the opposite of this sort of emotional bonding. Simmel's observations of the urban milieu found that the tempos of bodies, objects and things resulted in the individual figure of the *flâneur*. The *flâneur's* turn away from the outside to within was conjured by the nervous stimulation of the fluxes of modernity. But this is just one interpretation among many. Maffesoli (1996; see key idea box below) describes not the facing apart but an attunement between bodies to form what he calls 'the vitality of the masses' and the 'spontaneous intermeshing of warm, emotive human bodies into group formations' (1996: 34–6).

Key idea 4.4: THE ARABESQUE OF SOCIALITY

Recall if you will the discussion of the atomized subjects discussed in Chapter 2. The *flâneur* represented at once a fascination with the modern world as well as a distinctive exertion to remove oneself from it by paradoxically getting as close as possible. The *flâneur* and the contemporary nomad represent a breakdown of communities and social bonds through a self-conscious individualism (Sennett 1998).

In his polemic, *The Time of the Tribes*, sociologist Maffesoli rejects the atomization and individualizing tendencies of everyday life and urges his readers to consider the 'affective nebula' that ties people together. Part of this nebula is constructed by what he calls a *tactile relationship* that forms the materiality of *being together* reminiscent of our crowd theorists discussed earlier, particularly the ideas of Elias Canetti (1962). As Maffesoli writes, 'Within the mass, one runs across, bumps into and brushes against others; interaction is established, crystallizations and groups form' (1996: 73).

Maffesoli's tribes are characterized by their fluidity of occasional gatherings and dispersal, and he suggests examples can be found in the streets of modern cities as 'the amateurs of jogging, punk or retro fashions, preppies and street performers invite us on a travelling road show' (1996: 76). It is through the repetition of these acts that an aesthetic ambience is made that allows the 'instantaneous condensations' of bonds, which, he suggests, 'are fragile but for that very instant the object of significant emotional investment' (1996: 76).

In resembling the meaningful place-ballets understood by David Seamon and Jane Jacobs discussed earlier, Maffesoli argues that these movements are imbued with more-than-discursive significance, but emotional investment that may not be consciously attributed.

Dwelling on Jean Baudrillard's writings on the American highway and its traffic, Maffesoli recognizes the 'strange ritual and the "regularity of these flux(es) which put an end to individual destinies"'. For Maffesoli, Baudrillard seems to conclude that societal interaction or the 'warmth present' can only be found through the 'collective compulsion' of 'propulsion': mobility (this could resemble Baudelaire's discussion on traffic).

Maffesoli presents us with an almost animalistic or primeval social that sees individual mobilities and movements as part of a 'vast ballet' of which they have little consciousness. Bodies in their balletic dance bond tightly together into a system of constellations. Disposing of intentionality or reflection, 'neither will nor consciousness play a part', for Maffesoli, 'This is the arabesque of sociality' (1996: 76).

Further reading

(Canetti 1962; Baudrillard 1988; Maffesoli 1996)

The shifting of affective mood into gear could even be seen in public riots and in eruptive demonstrations or even in the contemporary formation of 'flash mobs' as forms of coordinated sociable activity. Kathleen Stewart (2007) explains how flash mobs emerge as swiftly forming public demonstrations organized by text messages or other means of communication. From her book *Ordinary Affects* Stewart recounts several recent examples:

> At a Toys 'R Us, a flash mob stared at an animatronic *Tyrannosaurus rex*, and then fell to the floor with screams and waving of hands before quickly dispersing. In New York, participants assembled at the food court in Grand Central Station, where organizers (identifiable by the copies of the New York Review of Books they were holding) gave mobbers printed instructions regarding what to do next. Shortly after 7 pm about two hundred people suddenly assembled on the mezzanine of the Grand Hyatt Hotel next to Grand Central Station, applauded loudly for fifteen seconds, then left.
>
> (2007: 67)

Events such as these illustrate the eruptive and mobile nature of affect. Flash mobs are ignited by the flare-up of 'little worlds, bad impulses, events alive with some kind of change' (2007: 68). Although the mobs may appear spontaneous they are usually highly organized, even though this does not detract from their performance which is 'exciting'.

Comparisons can be made between this and the kinds of participative mobilities that involve an experience of sharing. Like before, the experience of mobile physical involvement may be more than a representational or reflective moment where one deliberates what they have shared, but, for geographer McCormack, it is rather 'a presentation of movement'. Following Bateson's writings on presentational dance, McCormack attempts to take seriously how during the shared experiences of movement and dance, the body becomes more than 'me' or oneself, but external pathways that interrupt other bodies, moving through and over them (recall Zipf's [1949] conception of the body explored in Chapter 1). Quoting Bateson, the 'individual nexus of pathways which I call "me"' seems to be no longer so important or "precious"'. The nexus is part of a larger assemblage of minds and bodies (Bateson in McCormack 2002: 474).

Describing his experience at the Chisenhale Dance Space in London during 2001, McCormack recalls,

> We work around bodies, brushing against them, touching, watching them shuffle nervously back and forwards and up and down as they try to find space in and around our movements. Collective anxiety, connective affects.
>
> (McCormack 2002: 194)

By moving together affect appears to offer pathways of connection linking up the individual to an unconscious collective. Discussed in the case study box below, the affective nebula can emerge in other designated environments, such as a club.

Case study 4.4: THE AFFECTIVE MOVEMENTS OF CLUBBING

As mentioned already the space of the night-club is a prescient example of the formation of an affective collectivity. Through the intertwinement of bodies, music and movement, Ben Malbon's (1999) *Clubbing* illustrates how an emotional bond can be created uniting 'ostensibly disparate individuals'. Closely following Elias Canetti's theorizing of crowds and Maffesoli's analyses of tribal behaviour, it is the feeling of 'we' and the 'togetherness' that is part and parcel of the pleasure of dancing and moving together.

The music itself contains a power of identification that can link subjects through practices of listening and moving, which, Malbon argues, constitutes a special sensation of *extasis* or in-betweenness – a 'flux between identity and identification'. It is by mobility, by 'movement, proximity to, and, at times, the touching of others' that the individuals within a dancing crowd 'can slip between consciousness of self and consciousness of being part of something much larger' (1999: 74).

As will be more than familiar by now, the feelings and the processes of clubbing are quite difficult to articulate. Malbon argues that 'the relationship between movement and thought (or motion and emotion) is central, even if this relationship can be difficult to define or explain' (1999: 86).

Further reading

(Canetti 1962; Malbon 1999; Saldanha 2007)

We can upscale these sorts of bodily pathways to a more extensive kind of collectivity. An affective collective can be made on the move. Kevin Hetherington (2000b) takes the inter-personal communities formed by new-age travellers. Moving together as a group of travellers – the convoy – can enable senses of belonging providing 'participants with a sense of membership grounded in strong emotional experience and a shared sense of ordeal' (2000b: 78). Moving together as a community can construct an affective or emotional communion, similar to that of a pilgrimage (Bajc 2007; Bajc *et al.* 2007; Eade and Garbin 2007; Cavanaugh 2008). Experiencing the convoy together is expressed in terms like the 'vibe' and the 'buzz', or even the mood or tone of their journey. This appears to be the basis of an affectual identification of fellow feeling – a 'communitas of the pilgrimage *en-route*' (Hetherington 2000b: 75).

At an even wider scale, we can consider how these relations form through more complex choreographies of places. Taking the moment of arrival and departure of migrants into new places and environments, David Conradson and Alan Latham (2007) address the feeling of becoming-with or coming together through mobilities that produce new encounters and configurations of people and events. The *event* of the encounter is composed of 'lively' interactions forming ephemeral fields of affective charge, mobile 'configurations of energy and feeling' that rise and surge before sinking back and being reformed elsewhere. (Conradson and Latham 2007: 238). As they argue then, if it is 'not hard to think of identifiable places as offering specific affective possibilities', what happens when one moves to and encounters new geographical settings (2007: 237)? What are the affective possibilities for one entering, crossing and altering such fields?

A migrant's first encounter with a new city may be one of bombardment by wholly new 'sounds' and 'smells', different languages and vocabularies or dialects, and very different ways of doing things. Before a migrant might even have the chance to explain or describe what is going on, the authors show how the performance of these multiple mobilities may solicit all sorts of feelings and intuitive responses:

> At each moment of encounter with these new forms of urban life, but before an individual is necessarily able to explain what is happening or give it a name, affective responses – perhaps evident in the tightening of one's muscles, dilation of pupils and release of adrenaline – will be taking place. These responses might include surprise, joy, anger, or fear, sometimes in combination: a journey on a crowded London commuter train is certainly capable of eliciting all these affects in a short space of time.
>
> (Conradson and Latham 2007: 236)

CONCLUSION

How is mobility done? While this question might seem a little late to be addressing by now, it is reflective of the wider academic slowness to come to grips with the enactment of mobility. As we explored, the visual and the representational have often come at the expense of issues of practice, performance and the more sensual experiences of movement itself.

Cutting a swathe through diverse literatures on movement, representation and the body, the chapter explored how in composing a host of social processes and practices mobilities are far more than simple, conscious and calculative acts than we might have been led to believe in Chapter 2. Juxtaposing a range of conceptualizations of mobility from several different theoretical positions, mobility was understood as a multi-sensual activity which is not always consciously thought or representational. These issues, as we examined, were not merely the concern of high theory, but played an integral part in the formation and experience of social practices. They explain how mobilities are often habitual as shown in the case of driving when we find we have driven to our nearest gas station without thinking about it. They shed light on the way we feel when we are mobile; moving may enable us to think more clearly, or it can cause considerable pain. But this is not simply a question of representation and practice or the thought and the unthought. As shown, these issues serve to make us think more carefully about what thought and representations are.

Engaging in the more-than-representational emotional and affective dimensions of mobility enabled us to continue to move beyond the predominance of a singular mobile and a rational individual. Connection, collectivity, address and combination, all of these terms were strongly associated with the doing, practice or performance of mobility. Mobilities match up people and things. Being mobile-with seems to unlock barriers between bodies, enabling the passing on of ideas, emotions and fellow sentiments so that a feeling can become mobile itself. It is in this capacity that the book will consider how mobilities appear to enable a *mediation*.

5

MEDIATIONS

> [A]nything that moves must be carried in some way.
>
> (Abler, Adams and Gould 1971: 389)

INTRODUCTION

You have arrived at Toronto airport on your flight from Hong Kong. Upon arrival and departing the aircraft you are met by several security and immigration officials pointing a strange-looking device at you. The situation is even more disconcerting when you become aware of the large numbers of passengers and staff wearing cosmetic facial masks – as if about to perform surgery. This was the situation that confronted many passengers travelling both within and from South Asia in 2003. The SARS (Severe Acute Respiratory Syndrome) outbreak had sparked a security and health emergency. The virus spread through the movement of people travelling to and from China. From Singapore, to Canada and the Western United States, as well as Hong Kong, Taiwan and regions within mainland China, business travellers, tourists and migrants carried the disease which mimicked a form of pneumonia (www.openscar.com). Along this journey it had infected over 8,000 people claiming around 700 lives. Officers employed at the airport border were using a detection device that could observe raised passenger temperatures – an important symptom that signified a carrier of the virus.

We will revisit this example in more detail later in the chapter, but dwelling on the event now we might consider further the outbreak of public anxiety during those months. The face masks I mentioned were commonplace in Asian and Canadian border zones and airport terminals. Many people decided not to travel. News broadcasters made predictions and used computer simulations to illustrate the potential spread of the contagion. All sorts of mechanisms of governing and regulating people's movements kicked into gear. These were mechanisms that Keil and Ali describe as 'tied up with the regulation of bodies – human and nonhuman – that carry or are suspected to carry the disease' (Keil and Ali 2007: 853).

The example of SARS is an unnerving yet poignant example with which to start this chapter. The authors (Keil and Ali 2007) use the word 'carry' to describe the action of mobilities, be they bodies – human and non-human – or even satellite transmissions. These mobiles act as vessels and paths – conduits for the movement of something else. The mobilities of people meant the potential mobilization of the disease. SARS could spread incredibly quickly across the globe; it could spread within cities. It could be passed on within the aircraft cabin, and its news could be spread quickly by various media vectors. One mobility – the movement of people – meant that the mobility of another form of life became entirely possible. And people weren't the only mediators either. Passengers were mediated by aircraft and trains. Print and other forms of electronic communication transmitted news of the disease further across the globe. Mobilities carried other mobilities. Different kinds of mediating mobility such as an aircraft – which acted as a medium to a passenger body – could then act as a vehicle for the virus. Mobilities *mediated* the journeys of other mobiles.

This chapter will explore several facets to these mediations by dealing with several types of mobility that appear to have become conceptually separated from one another. Marshall McLuhan was perhaps the first to note how the term *communication* has gradually become divided from transportation, leading Thrift to later assert 'transport and communication cannot be split apart' (1990: 453). For McLuhan (1964) this division owes itself to the estrangement of informational mobility such as words, ideas and imagery, from physical objects and commodities such as stone, coins and papyrus. The divorcing of messages from these mediums meant the subsequent untying of roads, wagons and physical transport infrastructures from their fluidity. While the chapter will focus upon all sorts of electronic, informational or physical mobility that could be considered both communication or transportation, it is where different mobilities like communication and transportation cross over – how they mediate – that more compelling questions can be asked.

Thrift (1990) sums up this interplay explaining how transport and communications each rely, 'upon the other in all manner of ways'. We are told of the tightly coupled evolution of newspapers with the history of the Post Office 'along with the railways (or railroads) with the telegraph' (1990: 453). As Thrift goes on, mediation is about connection and complexity as well as the everyday. These mediating mobilities are ubiquitous in the way they are 'wrapped in and wrapped around people's lives' (1990: 453).

In order to explore these issues the chapter deals with several different kinds of mobile mediation. These are how:

- Mobilities are often diffused
- Mobilities mediate between
- Mobilities augment relations and other mobilities

First, the chapter simply demonstrates how societies are incredibly *mediated mobile societies* that permit, facilitate and enable social relations. Mediated travel has become so important that the mediator – the means of travel – have become objects of identification for national cultures, local identities and collectives of hobbyists. The chapter then moves on to discuss the process of *diffusion* as mediation – how mobilities may carry and transmit the mobility of other things. The chapter moves on to examine how in doing this kind of carrying, mobilities act as mediators *between*. What implications these go-betweeners have upon distance, our relationship with the environment, the landscape and each other, are investigated according to their capacities to connect and disconnect. Finally we will explore the properties of mobilities that are *augmented* by various technologies and objects that travel along with, track, enable and regulate passage and displacement.

MEDIATED MOBILE SOCIETIES: AUTO- AND AEROMOBILITIES

The automobile is often understood as the signature technology of contemporary society. Urry has regularly written that much of what 'people now think of as a "social life"' would just not be feasible without the use and flexibility of car travel. The car enables movements to be made at a whim, with little planning or preparation time so that one may travel to work, to family and friends (2000: 190). It is impossible to think about societies without contemplating how everyday activities are constituted by car travel. Various phrases such as Autopia and Car-culture have evolved in this literature (O'Connell 1998; Thomas *et al.* 1998; Miller 2001a; Wollen and Kerr 2002). Nevertheless, it is not only the car to be interpreted in this manner. Other transport technologies such as the aeroplane have instigated similar opinions. Media and cultural studies authors Gillian Fuller and Ross Harley (2004) use the wonderful term *aviopolis* to describe our increasingly *aero*mobile world. And of course transport scholars have suggested how '*mobility is a fundamental human activity and need*' and a fundamental 'touchstone' of the twenty-first century (Knowles and Hoyle 1996: 4).

Before we discuss the various processes through which mobilities mediate in any more detail, the section will begin to assess how mediated our societies really are. It is first just worth noting that the examination of transportation systems such as the car and the aeroplane, as you will have guessed by now, cannot be pinned down to one disciplinary approach but many. The following key idea box considers the sub-field of transport geography.

Key idea 5.1: TRANSPORT GEOGRAPHIES

As Goetz (2006: 231) has recently pointed out, 'transportation is central to the study of geography, just as geography is central to the study of transportation'. Providing in-depth analysis of phenomena such as urban transport systems, inter-state and inter-city networks, the hub-and-spoke systems of airport and airline networks, transport geography has become a popular subfield.

For some, transport geography hasn't quite kept up with the times, resembling and utilizing the sort of approaches described in Chapter 2. David Keeling's recent overview summarizes this kind of perspective through a review by Susan Hanson,

> transport geographers have not kept pace with theoretical and methodological advances in the discipline . . . transportation has

> "lost its disciplinary centrality, largely because it has remained within the analytical framework of the 1960s".
>
> (Keeling 2007: 218)

Others have asked very similar questions, 'Why did transport geography fail to keep up with the intellectual challenge provoked by such new thinking?' Hall and his co-authors enquire (Hall *et al.* 2006: 1402).

Despite lacking the apparent conceptual sophistication of human geography and the social sciences, perhaps this critique is a little unfair given that the discipline does provide its own useful and unique role. Others suggest that the contribution of transport geography to these debates has even been underplayed. Keeling goes further in his review drawing on Preston and Goetz, who explain how transport geographers have been able to draw out useful relations between geography's 'conceptual structures and empirical contexts' in recent times (Keeling 2007: 218).

Moves are afoot to respond to the 'new mobilities paradigm' in order to learn from its more nuanced understanding of transportation as a social object. For Keeling, 'transport shapes society and space in myriad practical ways. Yet society also shapes transportation in ways that are critical to the ordering of space, place, and people' (Keeling 2008: 277). There are challenges to this of course and Keeling devotes some thought as to whether 'moving beyond a fairly "static" social science approach to mobility will promote harmony or sow discord' (Keeling 2008: 277).

So far at least, the signs are positive. Knowles (Knowles *et al.* 2007) and his co-authors' reinvention of the classic textbook *Transport Geographies* is suggestive of a more holistic and integrated approach towards transport that sees mobility issues addressed by complimentary quantitative and statistical approaches alongside qualitative analysis.

Further reading

(Hoyle and Knowles 1998; Keeling 2007, 2008; Knowles *et al.* 2007)

The two main forms of mediation we can start with are probably the most hegemonic kinds of mediated mobility today: the transport systems of the car and the aeroplane. The technology of the car has received most discussion with many articles, monographs and special issues devoted to its analysis. The term 'automobility' has emerged as the dominant description of mobility mediated by the car. In Urry and Sheller's schema (2000: 739) the manifestation of the conjunction between machine and human creates a 'car-driver'. Thereafter, automobility means the new possibilities for mobility which emerge between bodies, cars and also the wider assemblages of 'roads, buildings, signs and entire cultures of mobility' (Sheller and Urry 2000: 739).

According to Urry, the car means social interaction occurs directly through this mobile augmentation, so that 'People dwell and socially interact via [this] movement' (Urry 2000: 190). In other words, automobility means more than the personal freedoms we saw tied up in the meaning of the car in Chapters 2 and 3 for, as discussed in Chapter 4, those freedoms enable new social proximities. The car and the aeroplane permit something much more than 'simply an extension of each individual' (2000: 190).

Aero- and automobility enable people to perform social activities as well as the obligations required to form and sustain relationships and networks: from going on holiday and performing leisure activities to meeting friends and family at short and long distances, and making one's way to work or to business meetings. Air travel creates what Claus Lassen (2006) describes as 'corridors of travel'. These are conduits that connect places together along which commuters and regular travellers may connect with their corporate meeting, conference or sales appointment, creating new forms of socio-economic spaces and connections. Businesses, state institutions and other organizational centres see networked connections form along routes between places: London to Geneva, Frankfurt to Oslo, and Paris to Stockholm made up around 45 per cent of all business air travel in Europe during 2002–2005 (Derudder *et al.* 2008). Reconstructed from Derudder *et al.*'s (2008) reading of the Association of European Airlines database, Figure 5.2 presents a hierarchy of these networks.

Auto- and aeromobility enable social practices to be conducted within the space of these technologies. Analysing this trend in a number of projects devoted to the relationship between driving and working, Eric Laurier (Laurier 2001, 2004) has shown how working practices usually related to desk-based employment were regularly carried out by respondents while driving. Activities included continual communication with the office by dictating letters to distant secretaries, returning calls to

Figure 5.1 Doing office work on the motorway

Source: Copyright © 2006 by E. Laurier, reprinted by permission of Sage Publications

Cities	Business travel
London and Geneva	3,281,117
Frankfurt and Oslo	2,026,604
Paris and Stockholm	2,019,845
Amsterdam and Düsseldorf	1,737,635
Copenhagen and Frankfurt	1,096,543
Munich and London	1,080,402
Stockholm and Zurich	863,045
Milan and Copenhagen	853,438
Vienna and Munich	764,851
Brussels and Vienna	763,111

Figure 5.2 European business connectivities

Source: Adapted from Derudder, Witlox, Faulconbridge and Beaverstock (2008, orig. table 5)

complaining clients, reading and shuffling between different paper documents, all while attempting to drive at the same time (Laurier 2004: 264).

Laurier and his team's use of ethnographic methods allow the researcher to get close to the point of practice or performance, permitting the respondent to do more than simply recount their actions as the researcher witnesses and records the bodily practices that compose working on the move. We may consider an excerpt from their ethnographic study describing the working activities of a respondent, Ally:

> Ally is travelling fairly rapidly along the motorway. . . . In syncopation with extendedly scanning the road ahead, she glances across and down at the in-tray of printed e-mails balanced on my lap. If she did not have an ethnographer in the car beside her today, the documents would be balanced on the passenger seat. With my assistance in sorting through a large pile, she selects two documents and places them on the steering wheel in front of her before making her phone call. Once the document is on the steering wheel Ally talks quietly through the documents saying who she has to phone next, what will be difficult in the phone call.
>
> (Laurier 2004: 266)

In this light, aero- and automobilities are important not only for the job they do in shuttling people backwards and forth but also because of the space they provide for work and other sorts of relations. The carving out of this space is essentially the construction of a *space for time* which can then be used as opposed to being 'wasted'. Driving and passengering shed valuable light onto the complex uses of 'travel time' in other public transport infrastructures such as rail travel (Lyons *et al.* 2007; Holley *et al.* 2008).

Flexibility

There is obviously a sense of flexibility and control in how one may manage and operate to work where and when they want and take hold of one's patterns, routines and times. Cars and aeroplanes also give some measure of control over one's surroundings in which sociability and available time may be managed. Urry suggests that the car has become 'home from home', 'a place to perform business, romance, family, friendship, crime and so on' (2000: 191), and a place reflective and performing of one's identity. Indeed, while public transport is seen to negate the benefits of the car's construction of some form of private space, the aircraft cabin is gradually introducing these variables if one is able to afford it.

Even as the car and the aeroplane have become environments over which their users and passengers have some form of control, exactly how they are used is enacted within other margins of (in)flexibility.

The terms 'system' or 'regime' have been posited at automobility particularly, in order to denote the kinds of infrastructures or assemblages of actors and technologies that free up the driver's autonomy (Bohm *et al.* 2006). Personal car travel is one of the most flexible forms of mediated mobility, relative to train travel, because of the way it enables one to take charge of activities. For Urry (2000) this produces 'a reflexive monitoring not of the social but of the self. People try to sustain 'coherent, yet continuously revised, biographical narratives' (191). Whereas the time-table of the railway creates a time for all travellers to embark and disembark (Schivelbusch 1986; Giddens 1990), the car creates the possibility for many 'times'. It allows 'personalized, subjective temporalities' for people to move at a time of their own choosing, to a destination of their inclination, without recourse to the rigidity of a system like the railway. The rapid growth of low-cost air travel and the use of private and corporate jet travel has further enabled the increasing flexibility of international travel.

On the other hand, while the systems of the automobile seem to have permitted greater flexibility, for other mediated mobilities, the reverse may be true. Within the air-travel industry, one could argue that room for flexibility has decreased given the incredibly rapid growth in the size and complexity of the airport–airline system (Urry 2007). Anyone who has travelled long-haul will be aware of the vast infrastructural complexities of international airport terminals that must convey huge numbers of passengers, baggage and cargo, such as London Heathrow, Chicago O'Hare, Schiphol Amsterdam or Changhi, Singapore (Graham 1995; Kellerman 2008). While these spaces were once very simple requiring little time to negotiate, today it is quite regular to get to an airport 4 hours before one's flight, in order to allow time to be checked-in, to go through security and to negotiate the shopping-mall-style environments (Wood 2003). Furthermore, the famous hub-and-spoke arrangements of airports have also meant that passengers must pass through quite restrictive corridors of airline flight networks in order to get to their destinations (Derudder *et al.* 2007). However, as the regulative rigidities of air travel appear to be lightening their touch through policies of de-regulation and liberalization, the point-to-point networks of low-cost airlines and regional airports appear to be loosening up a more flexible system (Dresner *et al.* 1996; Francis *et al.* 2003).

Inequality

In Chapter 3, we dwelt upon the exclusivity of mediated mobile infrastructures such as the road. Auto- and aeromobilities have been posited as particularly unequal mobile mediations. The mono-environments of the car dominate space and time 'transforming what can be seen, heard, smelt and even tasted' (Urry 2000: 193). Beckmann (2001) describes how a 'feet-only user' is barred from sites such as the drive-in cinemas and faces difficulties in navigating road-only environments. For those unable to afford the individual transportation of the car, 'the more he or she gets pushed onto the sidewalk' 'In such car-based cities, walking . . . has vanished from public conceptions of mobility' (Beckmann 2001: 598). Airports exert similar power over their local contexts by disturbing nearby geographies with material, economic, political and social changes. Airport infrastructures (Fuller and Harley 2004; Pascoe 2001) terraform their environments by shifting ground and earth. Their eventual completion constructs fields of sound and air pollution that repel, reorder and polarize local geographies.

Mediating mobilities have the capacity to modulate and preselect the people that use them almost as much as the landscapes they alter. Evidence suggests how the users of air travel are predominantly the wealthiest members of society. Even the rapid growth of low-cost air travel in Europe and the United States has failed to truly democratize it. The Civil Aviation Authority (2004) survey of 180,000 passengers at UK airports shows how many passengers at Stansted (London's low-cost 'hub') were from households with average incomes over £50,000. While enabling more people to fly who wouldn't have been able to afford it beforehand, low-cost airlines have enabled the increased frequency of flying for the same affluent people (Adey *et al.* 2007).

Attachment

Even though it has become knitted into societies and cultures that it is almost mundane, transportation is subject to interest and identification. In earlier times one used to talk of being air-minded – the belief in the promise of air travel (Adey 2006a). And of course the car has been a source of focus for hobbyists, collectors and enthusiasts. There are many reasons for the tying-up of social identities with these transport technologies, many of which fulfil political projects of corporate ambition, nation-building and citizenry.

In providing a window or a way for 'gazing' upon the nation, for Tim Edensor (2002), the cultural formations of both aviation and automobile enthusiasm have occurred through both symbolic representation as well as embodied forms of material practice. Historians of aviation provide glittering examples of the symbolic capital employed by the aeroplane in various societal contexts. Peter Fritzsche's (1992) stunning analysis shows the radical deployment of the aircraft in the German National Socialist Party's propaganda campaign for power and persuasion. Just as historian Robert Wohl (1994, 2005) demonstrates the dramatic visual power of the aeroplane moving through the sky to inform and represent various cultural movements in art and expression (Pascoe 2003).

Transportation academic Kevin Raguraman (1997) provides an example of the symbolic capital expressed by the 'flag-carrying' Singapore airlines. Perhaps more so than other means of travel, civil aviation acts as a symbol or an allegory for the nation as well as an efficient tool for nation building. Just recently in the United Kingdom the many problems that beset the first few weeks of London's Heathrow Terminal 5, were described by the county's newspapers as a 'national disgrace'. Just like airports, airlines act as faces for their nations, representing them back to their citizens and to those looking on from without. An airline prompts and promotes national consciousness. The symbolic power of the airline pulls people together. But as we saw in the previous chapter, mobilities are more than symbolic and representational. Identification and the constitution of national identity can occur in habitual performances of mobility such as driving (see box below).

Case study 5.1: SPACES OF INDIAN MOTORING

Tim Edensor's *Material Culture and National Identity* (2002) argues that the car is fused into the national consciousness by everyday and mundane forms of practice and doing. Drawing out the 'auto-centric' cultural practices which are, in turn, located within their own distinctive cultural spaces, we are presented with a rich example of Indian motoring.

The rules of the Indian road prescribe conventions and social norms (Merriman 2005a). These include how it 'is necessary to sound your horn to warn a vehicle that you wish to overtake, for rear-view mirrors are rarely used and often absent' (Edensor 2001: 120). Other norms require that road users comply to the rule that precedence is given to the largest vehicle. Cars give way to buses and lorries, just as rickshaws defer to cars, and bicycles may allow auto-rickshaws and motorbikes to overtake them. The road-scape is lined with temporary dwellings within which industries and services abound: 'bicycle-tyre repairer men and telephone kiosk wallahs, to roadside *dhabas* (small cafes and tea shops) alongside other services such as hairdressers, dentists and sellers of all kinds' (Edensor 2002: 120). The traffic of people and vehicles 'criss-crossing' the street in many directions disrupts the linear progress associated with Westernized road spaces. These include the 'miscellaneous collection; of people, things and animals, from carts, cattle, buses and other forms of transport, "providing a fluid choreography at variance"' to the apparent order and control of other Western roads (2002: 121). Finally, a distinct experience of Indian automobility is afforded by the very sensuality of the encounter itself, an encounter of 'continual manoeuvres: pressing the horn, jerking the wheel and applying the brake' (2002: 121). The scents of cooking, 'cow dung' and the fumes of the road impinge upon the olfactory. The soundscape has its own qualities that form an orchestral performance symphony of colliding 'human and animal noises' music and car horns.

Clearly these experiences and one's access to these road spaces are uneven and stratified. As a costly luxury, the majority experience the car not from within, but from the position of the side of the road, a seat on the bus driving past, or the bicycle and rickshaw weaving their way between busy traffic.

Further reading

(Dimendberg 1995; Edensor 2002, 2003; Merriman 2005b, 2006b)

According to the sensations and haptic experiences discussed in the previous chapter, all of these percepts add up to quite specific identifications with material transportation. Elsewhere the infrastructural spaces of other roads and airspaces have been imbued with powerful forms of attachment and significance, shown for instance in the reception of the M1 motorway in Great Britain (Merriman 2007) and the German autobahn (Dimendberg 1995), or in the airport terminal as travel writer Pico Iyer shows (2000).

DIFFUSION

So we have thought about just how mediated our societies are by the most visible, contentious and dominant forms of auto and aeromobilities. In this section we can take issue with the core component of mediation: how mobilities enact the action of carrying or being carried in order that something may be transmitted or passed on. This has been known as a process of 'diffusion'.

One of the most influential sets of writings on this issue refers to processes such as globalization. Polly Toynbee's (2000) analogy of a frothing, creeping and oozing giant strawberry milkshake, helps us to imagine the tidal wave of ideas, commodities, stuff and so much more, which have enabled Western culture to spill over the brim of its frontiers. The carriers of culture are identified by anthropologist Arjun Appadurai (1990) as various disjunctures between different mediums which are very mobile in and of themselves. Appadurai envisages ever-shifting mediating landscapes composed by the mobilities of 'people, machinery, money, images, and ideas' that 'follow increasingly non-isomorphic paths' (1990: 301). Appadurai (key idea box 5.2) shows how these mobilities are very often connected. It is because potent ideas and imaginations get passed on through commodities, such as film, that more mobilities and flows are created.

Taking the cultural flow of how films like Mira Nair's *India Cabaret* cross distance and borders, Appadurai comments on how young women travel to the city of Bombay because they are seduced by the representations of its 'metropolitan glitz' and come to work as cabaret dancers and prostitutes 'entertaining men in clubs with dance forms derived wholly from the prurient dance sequences of Hindi films' (1990: 301). The irony seems to be how these practices provoke further imaginations of such women with a 'looseness' engendering the real mobilities of men to come and visit them (1990: 301). Likewise, sex tourism in Thailand and Asia reinforce themselves in 'the economics of global trade' and 'brutal mobility fantasies' that traverse space and time (1990: 303).

Key idea 5.2: LANDSCAPES OF GLOBAL CULTURAL ECONOMIES

Looking in more detail at Appadurai's thesis reveals a complex and hierarchical set of 'scapes' that enables cultures to slip and slide over spatial borders and between the boundaries of economies, politics and societies. The first three scapes Appadurai constructs can be broadly found within three main fields: people, technology and capital.

The first, *ethnoscape*, really means people, or rather a landscape of moving people who 'constitute the shifting world in which we live'. These could be peoples such as 'tourists, immigrants, refugees, exiles, guestworkers and other moving groups and persons' (1990: 297). The second scape, the *technoscape*, refers to new landscapes and increasingly fluid configurations of technology. Finally, *financescapes*; here Appadurai is essentially talking about the ability for money to move, how the rolling conveyor belt of currency markets, stock exchanges and commodity speculations facilitates the movement of capital in order to oil the wheels of further mobilities.

These scapes are not independent; disjunctures tie them tightly together in unpredictable ways. So while India might export its labour of waiters and chauffeurs and recently builders to somewhere like Dubai, it might also send software engineers to the United States, where they become wealthy 'resident aliens'.

The above scapes enable two further ones: *mediascapes* and *ideoscapes*. Mediascapes consist of the landscapes of information produced by the far-reaching distribution of technological capacities to disseminate data. These include newspapers, magazines, television stations, film production studios and others (1990: 299).

Feedback loops appear as mediascapes provide the complex 'repertoires of images' and narratives of apparent 'ethnoscapes' to viewers far away (1990: 299). As we have seen, the passing on of these scapes, may in turn lead to the further desire and 'acquisition' of movement. Piggybacking these flows, other *ideoscapes* travel with considerable political import. These are the 'concatenation of ideas' of 'terms', 'images' and imaginations of, for example, 'welfare', 'rights', 'freedom' and others (1990: 299).

Further reading

(Appadurai 1990; Wark 1994)

Appadurai's exploration of cultural *diffusion* is by no means new. Some of the first work to deal with this issue started at the scale of the human body, whose mobilities have almost always involved carrying something from one place to another. 'Human portage' has been argued as perhaps the 'most universal as well as the most primitive means of transportation' (Vidal de la Blache 1965: 350) in order to shuttle and exchange goods and foodstuffs. Portage has been essential for societies to develop, enabling the survival of families and kinships, trade links and the forming of simple relationships.

Contagion

In addition to the weighty diffusion of materials and bulky objects, more ephemeral things can be passed on. Torsten Hägerstrand's pioneering work in time geography came after his research in the area of innovation movement. Taking an area of central Sweden, Hägerstrand explored how farmers and farming communities came to accept various agricultural innovations, including economic subsidies for grazing improvements, or even technological innovations in medical methods of vaccination (Hägerstrand 1967). Understood to move in waves, innovations would propagate across communities and areas of space containing townships and villages that would eventually reach saturation point. His approach was a more organized and quantitative analysis of diffusion which could be compared to Sauer's analysis of cultural diffusion (1952).

Innovation diffusion was a useful model to investigate the spread of ideas in a variety of different contexts such as advertising, rural communities and the manufacturing industry. While others set about examining its utility for the movement of cable television (Brown *et al.* 1974), firearms (Urlich 1970), anthropologists looked at the stretch and spread of the trade in goods and foodstuffs (Adams 1974). In some cases, innovation diffusion has been likened to a form of contagion (Brown 1968, 1981). This could be something like the action of 'a rumor running like wildfire through a student population, or perhaps a new farming technique moving through an agricultural area' (Abler *et al.* 1971: 390). The theory follows that ideas are communicated swiftly through social groups, friends and neighbours. As people tell other people the idea diffuses its way through a population.

Innovations in agricultural techniques and processes are simply passed on by face-to-face communication as people congregate in social gathering spots or travel to each other's home (Morrill 1970).

Contagion is not just a nice metaphor to apply to innovation, but it is an important physical process which is central to the spread and proliferation of disease across space. As we saw at the beginning of the chapter, the mobility of contagion or disease is an important way that other sorts of non-human life may be mobilized and spread across vast distances with considerable velocity and in complex and uncertain ways (Gatrell 2005). Examinations have come from various places including branches of epidemiology, social studies of medicine, health geographies and elsewhere. Indeed, the study of epidemics has been used to improve understandings of diffusion itself (Cliff *et al.* 1981; Meade *et al.* 2000; Cromley and McLafferty 2002).

Mobile people can be the mediators of the diffusion of disease. When thought about historically, the spread of disease across the globe by European imperialism mirrors the contemporary diffusion Toynbee and others locate in Western cultural imperialism. Particularly nomadic and mobile populations are often at risk of the spread of disease. The semi-nomadic rural peoples of Mongolia (Foggin *et al.* 2000) have been investigated for just this tendency. Nomad circulatory mobilities are seen to return to administrative centres in year-length intervals which increases the risk of the spread of contagion (Mocellin and Foggin 2008).

But to locate the primary transmitters of disease as people would be far too simplistic, rather it is people who mobilize and mediate other mobile carriers. In the outbreak of foot and mouth disease in the United Kingdom during 2001 several of the non-human channels through which the foot and mouth virus was able to travel have been identified. Sociologist John Law (2006) argues that the capacity for viruses to mutate allowed them to 'successfully parasite their way along other displacements and flows'. This meant that wind could literally transport the virus. The mobilities of the animals themselves could do this too. Other factors included whether animals mixed with the infected ones, whether the pasturage was shared, and finally the distribution of meat and meat products that circulated through and from the spaces of infection (Law 2006: 3–4).

Returning to the issue of the SARS virus the chapter began with, Ali and Keil (2007, 2008) suggest that the increasing material connectivity of Toronto and Hong Kong created a vast number of global route-ways and conduits for SARS to spread. In this context, the topological geometry of the *actor network* (Latour 1993) provides an appropriate way to visualize these connectivities. Objects, cats, people, air-conditioning, spaces, all these acted as seguways for the passage of the disease (Keil and Ali 2007: 849). The brushing-up of bodies permits the spread of more than feeling as Mick Dillon and Luis Lobo-Guerrero (Dillon 2007; Dillon and Lobo-Guerrero 2008) show how the organic life of diseases such as AIDS and Asian Bird Flu makes bodies, organisms and things seem to line up into vectors for transmission (see also Braun 2007). The mobilities of the organic and the inorganic, the animal and the human are 'coimplicated' (Dillon 2007) in ways that turn their 'contingency, complexity, and circulation' into an imminent societal threat (Hinchliffe and Bingham 2008). Melinda Cooper (2006) renders a new world of biological insecurity (see Chyba 2002) made unsafe by the complex mobilities of disease (Elbe 2005, 2008). Describing the many dimensions of threatening biological movements, Cooper summarizes how the twenty-first century has been defined by these diffusions, how

> new pathogens were crossing borders that were supposed to be impenetrable, including frontiers between species (Mad Cow and Creutzfeldt-Jakob's disease); contagions were hitching a ride on the vectors of free trade (the deregulated blood market that enabled the contaminated blood scandals to happen); the complex cross-border movement of food implicated in Mad Cow disease, perhaps even on the mobile vectors involved in the production of transgenic crops and therapeutics.
>
> (Cooper 2006: 115)

But what are the social implications of these sorts of mobilities? It is obvious that there are numerous problems and complex implications for international security and different modes of governing and securitization across territories and between bodies and populations (Foucault 2007). Right at the beginning of this chapter we were presented with the stringent border controls at the Canadian airport set up as a reaction to the SARS outbreak in South and East Asia. The potential for disease to be mediated by mobilities has led to the construction of numerous forms of biopolitical governance and management in order to filter and keep these stowaways out of cities and borders. According to Keil and Ali (2007: 853), there are consequences for 'governance as the carefully guarded distinctions between people and germs . . . are now being challenged unexpectedly'. As they suggest, these challenges have led to defensible policies and practices of regulation that aim to filter and restrict people, bodies and populations.

According to John Law, the defensive strategies aimed at regulating mobility have occasionally made the problems of mediated movements worse. Drawing again on the context of foot and mouth disease in the United Kingdom, Law shows how a culture of safety arose from the earlier problems of BSE in the late 1980s. Greater restrictions on food and hygiene safety meant far tighter regulation of sites and spaces through which meat was slaughtered and treated. Stricter regulation actually increased the mobility of meat to registered abattoirs, potentially contributing 'to the size of the foot and mouth epidemic' (Law 2006: 10–11).

Right as we contemplate the social consequences of the diffusion of disease and other mobilities, various scholarship is addressing the effects upon the *pre-social* through the transmission not of ideas or disease, but the sorts of affective feelings discussed in the previous chapter. In this instance, however, emotions and affects become mobile in themselves as they are transmitted between things – a kind of affective contagion. In the space of the classroom, Elspeth Probyn (2004) reveals how 'Bad jokes, shared laughter and a complicity between teacher and students, and amongst the students themselves, allows for the contagion of the interest-excitement affect' (Probyn 2004: 36). Media theorist Anne Gibbs locates the epidemiology of affect as something spread by the mediating technology of the television. Television acts as an 'intermediary' so that the feeling of a film or a party political broadcast can mean 'affect migrates from body to body' (Gibbs 2001). A cauldron of bubbling emotions, war can activate mobilities and feelings in ways that may drive populations into a panicked and de-humanized state (Jones *et al.* 2004, 2006; Bourke 2005).

From other perspectives the diffusion of feeling works more like that of physical contagion itself. Teresa Brennan's (2003) research explains how biologists have discovered pheromones and ectohormones that are emitted by humans and absorbed by others. The mood a room gives out 'adds' something to a subject by physiologically producing a new mood or an affect. Affects are mediated along different paths of communication. The new mobile medias of

Appadurai's technoscapes introduce for Gibbs (2001) what is 'a powerful new element into this state of affairs'. Televisual communication (Parks 2005) is able to heighten affects 'by amplifying the tone, timbre and pitch' of voices and faces while dramatically increasing its speed and reach. Of course, other cultural practices can form different channels for these movements. As we saw in the previous chapter, affects can be passed on through rhythmical body movements. Moving together along to music can enable forms of inter-human communication that express 'collective messages of affective and corporeal identity' (Cohen 1995: 443).

Mobile method 5.1: TELEVISUAL MOBILITIES AND THE MOBILE GAZE

What does television, film and cinema – or what Lisa Parks (2005) refers to as the *televisual* – have to do with the study of mobility? As we have already seen, the tourist, the consumer and even the train passenger have often been investigated as an act of cinematic voyeurism. In the context of train travel, Schivelbusch described this as 'panoramic perception' that mirrored the animated image of the cinematic screen (Schivelbusch 1986; Kirby 1997). Both cinema and physical travel can produce the effect of a 'mobile gaze', 'the images on the screen move, as does the camera' (Cresswell and Dixon 2002: 5).

But how might television, cinema and film be taken as an object of mobilities research and as a focus of methodological enquiry? Tim Cresswell and Deborah Dixon's (2002) intervention in this debate suggests that films can be studied for the way they enable their viewer to travel: 'As we watch the film we travel – we become somewhere else' (5). Following in the footsteps of authors from cultural, media and communications research, film can be constructed as some sort of tourism where we come to occupy distant or different places, and actually, different subject positions.

Film and television travels too, a point Appadurai demonstrates in his commentary on *India Cabaret*. By moving along the pathways of cinematic distribution, films may pass on specific cultural messages and values to receptive audiences. In this sense, the televisual can be examined as a mediator to the mobility of cultural ideas, which when facilitated by satellite communication and Internet broadcasts, may move in incredibly extensive ways.

The content of film and television can be looked at closer still, to uncover what Parks (2005), following Raymond Williams, defines as 'visual mobility'. Reminiscent of the camera moves Cresswell and Dixon referred to above, the movement of a positioned gaze can be an important cultural and political device to communicate feelings and ideologies. Thus 'visual mobility' can play an important role in popular textual methods such as discourse analysis.

Returning to the material and physical activities of watching itself, televisual experiences are constituted through (im)mobile bodily practices of watching, listening and paying attention. Far from focusing purely upon the image, researchers may attune to the consumption of the televisual in streets and airports as Anna McCarthy (2001) does, telling us useful things about the utilization and management of public and private spheres.

In practice

- Film and television can be discursively read for their transmission of cultural values, meanings and ideologies
- The viewer may be seen to undergo a brand of virtual travel as one is transported to other imaginary geographies
- Film and television can be investigated for their distinctive experience of mobile images, sounds, sensations and collective practices of watching

Further reading

(Williams 1974; McCarthy 2001; Cresswell and Dixon 2002; Parks 2005)

Music

Dwelling on music may help us to consider sound as another kind of diffusing and mediating mobility. Gibson and Connell (2003: 9) encourage us to imagine music 'at its most basic' as simply 'sound transmitted from the microlevel (in a bedroom, pub, car, between headphones) to the macroscale (through various means, including the global media)'. At its most fundamental form, music is essentially a material transmission or a movement through space that mediates a pattern of what we may or may not call music (depending on one's taste). Therefore, the mobilities of music are more than the mere physical mechanics of wave motion through air in more ways than one. Music as 'an artifact moving with people' is able to cross borders and distances and may be passed on as 'indigenous knowledge, oral traditions or records' (Gibson and Connell 2003: 9).

Geographer Tariq Jazeel argues that 'Sounds evade fixity and easy definition,' and because of this fluidity, 'they are difficult to draw boundaries around' (Jazeel 2005: 237). Music is extremely difficult to pin down and locate from a particular place or cultural origin because it moves along the paths of travellers, migrants, communications and commodities (Leyshon *et al.* 1995: 430). We could easily consider how music is ridiculously mobile as it can be bought, sold, tried, tested and transmitted across several different mediums, forms of transportation and communication. But even as it isn't too hard to imagine the diffusion of music by electronic mp3s, and other physical portable formats like CDs, tapes and records, we must not forget that 'other forms of music diffusion are based largely on the movements of people rather than products or capital' (Gibson and Connell 2003, 160). As Kevin Robins notes, 'With mobility, comes encounter' (2000: 196) and with the movement of people and the passing on of music, cultural forms may be transmitted.

Music carries considerable freight by transporting the baggage of different cultures. Ideas of 'home', 'locality' 'bounded identity' are transmitted forth into the different environment music might come to occupy. This is not to say that music simply overwrites the cultural spaces it comes to occupy. Musical cultures have actually evolved *because* of the movement of people and music. Taking the example of rap, Paul Gilroy's (1993) *Black Atlantic* traced the cultural pathways rap has moved and been transformed through. Gilroy questions the generalized 'assumption that to place rap is to explain it', for it risks 'denying the mobility, mutability and global mediation' of rap as a musical cultural form (Leyshon, Matless *et al.* 1995: 429). Gilroy's study essentially shows how musical cultures are rooted and formed in their own diffusions (Bhabha 1994). Many musical cultures have been born out not of fixity but mobility. The cultural displacement of triangular trading, European imperialism to South America or the 'traffic of economic and political refugees within and outside the Third World' (Bhaba cited in

Gibson and Connell 2003: 187) are prime examples of the role mobility plays in the emergence of new musical traditions, such as soul, blues and jazz in the United States. As new spaces and cultures are crossed by human mobilities, they are simultaneously paralleled in the musical cultures that form through these movements.

At the same time as we emphasize the incredible movements of music, immobilities are still really important to music, especially the way it is marketed and consumed. For Gibson and Connell, the metaphors of flow and fluidity, 'explain only part of the story' (2003: 46) for mobility actually 'triggers new attempts at fixity'. If one considers the importance of how music is marketed, location and origin are vitally important; they are often intensely important to the pieces of music themselves. Music is marketed through place, while places are marketed through music. Using its heritage as the home of the Beatles and as a point of cultural regeneration, Liverpool has marketed itself as a cultural hub for music (Cohen 2005).

MEDIATING BETWEEN

Anthropologist Tim Ingold (2004) suggests how the way one moves through the world is almost always mediated or altered by technologies and techniques of *footwork*. For Ingold, our perceptions and knowledge of the world and its environment occur through and with various technological prosthesis we use in order to enhance or help our journey along the way. These could include examples such as 'skis, skates and snowshoes; running shoes and football boots; stirrups and pedals; and of course the flippers of the underwater diver' (2004: 331), or other devices, like 'walking sticks, crutches, and the oars of the rowing boat' (Ingold 2004: 331). The environment is perceived through one's embodied mobility, which in turn is mediated by objects that can enhance, change or dilute the visceral experience.

Following Ingold's cue, this part of the chapter continues to take issue with two main themes. Firstly, in Ingold's terms, mobilities repeatedly occur alongside various technologies, objects and items that alter our *footwork* – the materials and actors which mediate between. Second, Ingold's paper alerts us to another mediating capacity of mobility; in addition to travelling with these objects, mobile prosthesis may act as the mediators themselves. Mobilities involve not only a mediation but a mediation between. We have thought about how mobilities can carry. We have considered the ubiquity of transport technologies in contemporary society, but we must also consider how mobilities get between other mobile things, places and peoples. Take the simple instance of carrying something like an object such as a briefcase. The briefcase is something carried, it is brought with one, it is something that could hold important documents and information – a laptop computer maybe. In carrying and moving with the briefcase its user becomes a 'prosthetic' subject (Lury 1997). The prosthesis fundamentally alters its user's orientation and experience in the world. The briefcase might make running for the train difficult. Its bulky appearance might make entrance through a narrow barrier a nuisance. On the other hand, the briefcase can enable connection. It might supply one with symbolic capital and prestige. It could spark conversation on a flight from someone with a matching case. And the laptop computer it carries could permit telecommunications on the Internet.

Just as the briefcase enables both physical and social (dis)connections, in this section we will explore how – by getting between – mobilities are able to bridge social, spatial and temporal divisions or, conversely, they may work to push them further apart. Discussed in the box below, Michel Serres conceptualizes these relations through two key figures: the angel and the parasite.

Key idea 5.3: ANGELS AND PARASITES

An important interlocutor in this debate, philosopher Michel Serres' conceptualization of messages are thought provoking and useful. One of Serres' (1995a) now famous figures is that of the angel. By angels Serres doesn't necessarily mean angels of the spiritual otherworldly kind, but instead it is about how mobile entities are able to distribute messages. By angels, Serres means messengers, entities with the capacity to pass on communications. 'Look at the sky,' Serres tells us, 'even right here above us. It's traversed by planes, satellites, electromagnetic waves from television, radio, fax, electronic mail.' For Serres the plethora of communications immerse us in celestial fields of information. Angels, therefore, represent 'the systems of mailmen, of transmissions in the act of passing' (Serres and Latour 1995: 118–119).

But on the other hand, the angel is matched against Serres' other figure, the parasite (Serres 1982). The things that mediate our mobilities, or the mobilities that mediate, are not necessarily so conducive to communication. The parasite engenders not so much transmission, but by getting between it creates noise and interference. It does not improve things or clear pathways for communication, but it can block and clog things up. Both figures, the angel and the parasite, present opposite ends of Serres' equation.

Further reading

(Serres 1982, 1995a,b; Serres and Latour 1995)

Serres' ideas set up lots of questions relevant to this section. They demand that we ask how certain kinds of mobility conjure up angels – messages that allow the free flow of information permitting all sorts of connection, proximity and communion. On the other hand, taking the parasite, we can think about how the mediator can send too many of its own messages and get in the way of smooth communication.

Connection

Perhaps the most obvious place to start a focus upon the connecting properties of mobile mediations is with the now familiar term 'time–space compression'. What writers such as David Harvey (1989) have put down to the effects of mobility technologies such as the railway and the aeroplane, the ability to mediate one's journey across time and space is seen as the key ingredient in the 'compression' of space. Growth of the nineteenth-century railway network along with other mediating technologies such as the telegraph, shipping, radio, x-ray, cinema, bicycle, automobile and aeroplane 'established the material foundation' (Kern 2003/1983) that enabled one to speed their body and information across space, while altering one's sense of it as they did so.

The sophistication these technologies imply diverts our attention away from the more simple inventions that dramatically transformed mediated mobilities. At its most simple, Paul Virilio (2005) demonstrates how the invention of the horse-as-transport enabled the ability to bridge between places, bringing points on the map much closer together. The body of the horse becomes significant as a 'body-bridge'. The ability to disperse spatially sees the mount apprehended as an 'elevated cross roads', literally an '*interchange*' (2005: 51). The wheel and the stirrup have been taken as other vital inventions. In their combination with people, other animals and machines, they unleash human movement (Mumford 1964; Toynbee 1977).

These processes of time–space compression have been given other names such as 'time–space convergence' – the incremental rate at which transportation has continued to bring places together by gradually shortening the amount of time it takes to get between them (Janelle 1968; 1969). By calculating these convergence rates Janelle found various steps, jumps and leaps various innovations had brought to transportation speed. Just what this process (Janelle 1973) has meant for the human capacity to act and effect was taken up by media theorist Marshall McLuhan (1964). Where William Mitchell would later argue that 'we are not fully contained within our skins' (Mitchell 2004: 38) making us 'spatially and temporally indefinite entities' (2004: 38) today, McLuhan (1964) had already predicted this phenomenon more than forty years earlier. Seeing a new and extended human being (see Janelle 1973 on extension), McLuhan saw how communications and mobility technologies such as the road or the newspaper, could expand people's abilities to sense and to receive and transmit information – constructing an expanded and prosthetic nervous system.

Janelle's time–space convergence and Harvey's compression describe a state of affairs where spaces seem squeezed much closer together. It may now make more sense to consider mobilities in terms of the time it takes to get from one place to another as opposed to miles or kilometres. Pushing places together in this way seems to have had a variety of implications for human relationships (Meyrowitz 1985; Allen and Hamnet 1995). As Harvey puts it, 'The experience of time–space compression is challenging, exciting, stressful, and sometimes deeply troubling, capable of sparking, therefore, a diversity of social, cultural and political responses' (1989: 240). We will discuss the more negative consequences in a few moments, for in many respects the apparent consequences of time–space compression cannot be understood without regard to the significantly positive meanings accumulated in efforts to overcome distance. The sorts of mediators Kern and others discuss 'have been of immense significance in the history of capitalism' (Harvey 1989:

232) given the 'considerable pressure to accelerate the velocity of circulation of capital, because to do so is to increase the sum of values produced and rate of profit' (Harvey 1989: 86). Understood as symbols of progressive thinking, mobile mediators have often been pursued and purveyed as liberatory technologies. Drawing on Wolfgang Schivelbusch's (1986) history of the railroad, Thrift (1996) explores how the mobilities of circulation are regarded as 'healthy' and 'progressive'. The idea that 'communication, exchange and motion' have overcome the obstacles of 'isolation and disconnection' to bring 'enlightenment and progress' (Thrift 1996, 200) is an old idea indeed. Not only have mobility and communication brought social relations together, but they have also stretched across ever-increasing distances. Gidden's (1990) calls this time–space distanciation, or the ability for social obligations, ties and networks to occur over a considerable amount of space. The many potential *connections* Urry (2002) explores, therefore, require some kind of spatial mobility which becomes both 'the prerequisite for, and the consequence of, social interaction.' (Beckmann 2001: 597). These could include all sorts of practices and new means of communication that we have discussed already so far.

While mediated mobilities have enabled what look like positive social connections and ties, it really depends on which way you look at them. The global unity supposedly created by the technology of the aeroplane came with a simultaneous feeling of anxiety (Kaplan 2006). The increasing possibility of proximity meant the consequent feeling of 'apprehension that the neighbours were seen as getting a bit too close' (Kern in Harvey 1989: 270). International security has long been threatened by the convergence of places and the increasing ability for a sovereign power to reach over spaces and territories. We saw this in Mackinder's fixation upon the mobile power of horses, ships and railways. The Cold-War era of inter-continental ballistic missiles was premised upon similar such debates (Der Derian 1990), while our concurrent age of terrorism has constructed ordinary people as potential mediators who could deliver a bomb in any place and at any time. Cultural and political theorist Jeremy Packer (2006) writes that 'we are all becoming bombs. . . . The war is on and whether you like it or not (upon entering the US Homeland), you are becoming a bomb.' Everyday movements take on the status of a 'threat vector' (Adey 2004) as mobility – 'the crossing of terrain' – has become a 'prime form of weaponry' (Packer 2006: 378).

The sorts of connections discussed so far emerge through the bridging of predominantly large-scale geographical expanses. Mediating technologies may join up disparate points on the globe, but we have said little to the more intimate bodily relations connected by mediating mobilities – particularly the bodies that have become indivisible from machines and technologies. It is impossible to see the human subject and machinic technology as particularly separate things (Thrift 1996: 112). The regulatory obligations of 'mirror, signal, manoeuvre' (Merriman 2006a) become more-than-prescribed driving practices, but techniques that occur with and through the technological machinery of an assemblage like the automobile. Accepting the mediated body through the figure of the cyborg (Haraway 1991; Lupton 1999), the mobile body suddenly becomes a mobile 'prosthetic subject' (Lury 1997), whose mobility may be enhanced, improved and sped or perhaps slowed, encumbered and frustrated as bodies travel with and in assemblages of technologies and objects .

Often taken as the paradigmatic cyborgic construction of the twentieth century, the 'driver-car,' as Tim Dant calls it (2004), has been interpreted as an assemblage of flesh, metal, wires and rubber. The driver-car offers a distinctive communication between body and space. The driver's practices use small-scale bodily movements and actions, and they are able to feel 'the bumps on the road as contacts with his or her body'. One may sway 'around curves as if the shifting of his or her weight will make a difference in the car's trajectory' (Urry 2000: 32). Explored in the box below, the minimal baggage of the skateboard permits a similarly augmented connection with the ground underneath.

Case study 5.2: SKATEBOARDING AND MEDIATION

Ian Borden's *Skateboarding and the City* (2001) is a benchmark text for our understanding of the mobile bodily negotiation of public architectures and forms. Borden envisages an active co-construction of body and space reminiscent of Ingold's phenomenology and more-than-representational styles of thinking (see the previous chapter). Body, tool and architecture combine into an event that always reproduces itself. The skater's co-construction occurs through performances that are witnessed by other skaters and watchers (2001: 124).

The collaboration of body and board is performed through the skateboarder practising sympathetic micro-bodily movements with the board. The feeling is not that the skateboard separates its user from the landscape around him or her. Through the mediation of the skateboard the skater seems to somehow gain a very different and yet more complete and direct interaction with the space around. 'You must get the feeling that your mind is located in the center of gravity. You must think and act from your center,' says one of Borden's respondents. This is a particularly direct engagement with space; the enrolment of body, board, movement and terrain are enveloped within one another. The skateboard is at once external to one's body, while it is 'absorbed within the dynamism of the skater's move'. The move's envelopment means the board becomes 'a mediation and tool necessary to the skater's relation to the terrain underfloor' (2001: 125).

The board unlocks the potential to engage with space in different and spectacular ways. Like the pedestrian becoming a practitioner of parkour, space is 'brought to life' as 'verticals', 'curves' and 'symmetries' are enlivened from their previously static and deadened state. Curbs, rails and steps become something rather different.

Further reading

(Borden 2001; Woolley and Johns 2001; Macdonald 2005; Nemeth 2006)

These mediations work to improve the conductivity between bodies and other bodies. Katz reminds us of how the driver interacts with other drivers through the small and micro-gestural movement of 'loosening and tightening the grip on the steering wheel' (Katz 1999: 32). Emotions and affects may be shared or provoked. Drivers continuously awaken other drivers from apparent isolation through a simple nudge of their presence; the beep of a horn and a finger gesture, or it could be by the touch of a fender-bender and an apparently selfish manoeuvre. 'No longer cocooned in our secure world', writes Deborah Lupton, 'separate and autonomous, we become drawn into hostile relations with others' (Lupton 1999: 70). The driver-car is, therefore, subjected to the traffic of the outside that it cannot help but feel. Latham and McCormack cite from Lerup's evocative description of this very liquid space in which

> the dance and the dancer are fused in a swirling, self-engendering motion promoted by the darting of the driver's eyes, touching (because so intimate, so familiar) street, canopy, house, adjacent car, red light, side street, radio station Tejano 106.5, car upon car, instruments, tree trunks, joggers, barking dog, drifting leaves, large welt and dip, patch of sunlight.
>
> (Lerup cited in Latham and McCormack 2004: 713)

Mediated mobilities like the driver-car can be about creating a space of one's own. They are about space for moving and moving for space. Driving to work can be a space for 'contemplation' enabled 'by an interlude in which no decisions about how to fill in time are required' (Edensor 2003). Edensor explains how the smoothness of the drive can cultivate thought, the forward planning of future encounters, or perhaps reliving prior ones.

Disconnection

We have seen how mobilities can mediate and be mediated in a way that facilitates certain sorts of connection; they could be a bridging between people and locations perhaps, or between bodies and environments. In these instances, the mediation of and by mobilities *connects*. On the other hand, this can work the other way: mediators can *disconnect*. Before we look at this in any detail, take one of the most common processes of disconnection you might not immediately think about: travel sickness. From psychology, James Reason's fascinating *Man in Motion* investigates the inconvenience that arises 'when two vital psychological processes are disturbed by the unnatural circumstances of passive motion' (Reason 1974: 6). The presence of motion sickness tells us things about the communication of messages between body and world.

The body's reaction and adaptation to its movement appear to depend on the vestibular system that encompasses the sense organs of inner canals and other receptor systems that receive and sense bodily mobility. According to Reason, for most people these kind of receptors are able to sense and register self-propelled mobility very effectively. They enable the body to react to its movement so that something like vision is able to compensate its focus if one's body moves closer or further away from an object. Mediated mobilities disrupt and cause conflict for these honed kinaesthetic receptors. Travel sickness is something many of us may have felt riding in a car, a bus or particularly a boat.

Reason gives us the example of a man standing at the side of a ship looking down at the waves (1974: 28). The ship's movement creates a disconnect between the two sorts of motion-information being sent to the brain, from the eyes looking at the waves, and the organs of balance and inertial receptors processing the ship's movement. Such a disconnect may serve to remove the subject from the space of the boat and their fellow travellers as they absent themselves to the bathroom! This example not only tells us something about the body's kinaesthetic dealings with motion, but it also shows us – in micro-bodily form – the importance of how mediation can confuse or cause conflict in messages.

In other contexts mediating mobiles may deaden the transmission of information. Michael's (2000) analysis of walking elucidates Serres' idea of the parasite through the peculiar example of the walking boot. Michael considers how boots are not necessarily about facilitating communication between foot and mountain. Rather the boot constitutes a kind of 'intervention' between what 'should be' an easy flow of information between the body and mountain: '[T]hey disrupt, abbreviate [and] curtail the signals' (Michael 2000: 115–116) that move between mountain and foot acting as a noise or a disruption between these flows. Likewise, Ingold points out that boots and shoes 'imprison the foot, constricting its freedom of movement and blunting its sense of touch' (2004: 319). Grips transform how one treads. A worn and hole-ridden pair may force its wearer to avoid puddles and boggy terrain and particularly sharp or rough patternings of ground (Michael 2000: 116).

For all the buffering shoes and boots achieve, a long history of writers and thinkers consider walking to be the only true and authentic engagement with nature, the environment and landscape. Others remind us how the horse-as-mediator distances one from the landscape because it is an inflexible arrangement. Walking is pleasurable because it is autonomous; one *sees* what they want to see while on the horse they ride by. Thus, following Rousseau's observations,

> You leave when you want, stop at will, do as much or as little exercise as you want. . . . If I notice a river, I coast by it. A thicket? I go under its shade. A grotto? I visit it. A quarry? I examine the stone. Wherever it pleases me, I stay.
>
> (Van den Abbeele 1992: 111)

Such perspectives should be balanced with those who explain how riding a horse or travelling with a donkey can offer different and sometimes much more attuned engagements with the environment (Game 2001; Merrifield 2008).

Although the car may form some kind of connective tissue between driver and landscape, it can act as a complete buffer between driver, landscape and other drivers (see the key idea box that follows). Margaret Morse's (1990) popular piece on the freeway offers up imaginary spaces more akin to the cinema or the novel. Like the abstract nodes, lines and points explored in Chapter 2, the freeway is just that. For Morse, the freeway is not only a 'waste of time spent in between, usually alone and isolated within an iron bubble' but it is also an 'intensely private space, lifted out of the social world' (1990: 199). The almost frictionless travel of the freeway enables a 'partial loss of touch with the here and now', forming a distraction as if watching a movie; car travel along the freeway thus constitutes a type of virtual travel that sinks the driver 'into another world'.

Key idea 5.4: THE POLITICS OF COMFORT

Paul Virilio's *Negative Horizon* (2005) gives us an amazing commentary on the emergence of mass passenger travel and its relationship with the comfort of the passenger body. With speed came the increasing mediation of the passenger by what he calls 'corporeal "packaging"' – how the traveller is 'squeezed into his upholstered mantle, in the arms of his armchair, and image of a *body mummified that moves*'. You will probably be familiar with your own packaging-up as you are wedged into the seat of the bus, enwrapped by the comforts of your own car, or strapped into your seat on the plane bordered by arm rests on either side (Bissell 2008).

These developments serve to protect the traveller, while they constitute a politics of comfort expressive of the aesthetic tastes of the wealthy and the privileged.

Travel in this way becomes more and more weightless. The vulgarities of the haptic senses of touch are deprived as the traveller loses more than these touches but the 'physical reality' itself, of having an actual body. For Virilio the passenger is, therefore, doped and duped; they become addicts to a passage that is smoothed out and manipulated. The passenger is lied to by mediators that shield them from 'muscular contact' and give 'way instead to a series of caresses, light strokes and fleeting slidings' (2005: 55; Schivelbusch 1988).

All of these add up to what Virilio sees as an imposition – an unwelcome visitor of 'mediating elements destined to cause us to lose complete contact with primary materials'. The passenger is left completely unable to find their bearings or attain any sense of position of where they are (2005: 55). Furthermore, the passenger is distanced from the kinds of sensations and feelings they might actually want (Adey 2007). Car drivers may seek out cars which provide a more sensorial and embodied contact with the elements (Merriman 2007), whereas others may explore further kinds of risky or fully embodied modes of travel and adventure (Thrift 2000b).

Further reading

(Virilio 2005; see Merriman 2007 for a critique; Bissell 2008)

On the other hand, the co-identification of comfort with authenticity has seen driving become more closely wrapped up with the body because of numerous intermediators. Driver judgements are being imposed on, managed and augmented by systems such as traction control, anti-locking brakes and more sophisticated forms of software management that fine tune the driver experience. These are seen to produce the paradox of much more heavily mediated inter-action between the body and road and, at the same time, perhaps a better, exact or more plausibly heightened contact (Thrift 2004a: 51; Dodge and Kitchin 2007). In some senses, the more mediated the car seems to be, the more invisible it becomes.

Although we might presume how mediated mobilities can facilitate the dislocation of people from places, there is a simultaneous withdrawal of people from the social spaces of mediated travel. Ervin Goffman's (1961, 1963) renowned sociological analyses of public space describes the contemporary problem of navigating chance encounters and involuntary contact within highly mediated environments. Explaining what happens when a few strangers come together in the enclosed space of a railway compartment, Goffman discovers the disposition to avoid other passenger's eyes and, particularly, to avoid staring. This seems to require the act of 'looking very pointedly in other directions' (Goffman 1963: 137). Such practices appear to be more than unconscious attitudes and habits but conscious acts which may express 'too vividly an incapacity or a distaste for engagement with those present' (1963: 137).

The vertical mobile technology of the elevator provides some of the most interesting examples of this trend which I explore in more detail in the case study box below. Most of us may recognize the sometimes quite awkward social encounter of when one is held temporarily immobile while waiting for an elevator and inside it. In the elevator itself a sociology of 'civil inattention' emerges. Inattention forms in passengers' over-attentiveness to 'the back of the operator's neck' and the 'little lights which flash the floors, as if the safety of the trip were dependent upon such deep concentration' (Goffman 1963: 138). The airport and aircraft cabin offer up similar places of *non-engagement*. For urban theorist Mark Gottdiener (2000) airports fulfil the role of a place bursting with non-engagement, inattention and the loss of a social scene. 'Departure lounges are not commonly a place for social communion,' Gottdiener notes (2000: 34), before going on to find many comparisons between the airport and Simmel's imagination of the city as an instrumental space *par exellence* where there is no need for people to interact. The 'airport norm is one of *non-interaction*,' writes Gottdiener (2000: 35), describing the collapse of people into their own cocoons: 'We do not bother others and they are not expected to bother us. When they do, the situation thus created becomes quite uncomfortable' (2000: 35). Passengers try hard to avoid 'the encounter with a chatty neighbor' (Gottdiener 2000: 35) by performing being-attentive to their magazine, their fingernails and other mundane tasks.

These sorts of practices counter many of the assumptions and mistakes regarding the apparent vitality of mobile bodies and subjects. David Bissell's (2007, 2008) series of contributions to this debate contradicts the association of mobility with productivity or indeed immobility with any sense of passivity. Bissell's phenomenology of the transportation spaces of train travel shows how an immobile body subject waiting within their train carriage does not conform to such simple dichotomies; travel and mediated mobility do not require simple excitable bodies. To be still requires effort. And the passenger experience may be one of an acquiescent withdrawal from strangers and their conversations. It could be a lull into submissive sleep by the rocking of the train (Harrison 2008; Kraftl and Horton 2008). Alternatively, the passenger may be immobile yet frustratingly active. The constant monitoring of information displays and one's luggage can be exhausting (Adey 2007, 2008). Passengers are frequently angered, agitated and incredibly stressed (Bissell 2007).

Case study 5.3: WAITING FOR THE LIFT

Hirschauer's (2005) fascinating study of elevator interaction examines how people perform 'being a stranger' by way of various conscious choices and unconscious dispositions in relation to social norms and values of how to behave. For Hirschauer, the vast mobilities of people in city spaces create quantities of 'insignificant others', strangers who clash in unpredictable encounters of *unrelatedness*. For Hirschauer, the narrow spaces of waiting such as an elevator problematize this unrelatedness as strangers must spend relatively more time with other strangers than they would normally do, and thus their unrelatedness needs to be 'interactively maintained'.

In his examination of the elevator Hirschauer suggests that the production and maintenance of strangeness cannot be accomplished by passivity, but by an active passivity wherein passengers *enact indifference* (2005: 59). These practices of indifference that can be labelled *inactivity* involve the avoidance of contact by looking away and glancing elsewhere, even when in a position of extreme proximity with other passengers. Such practices enable the denial of 'physical nearness as a sign of personal relationships'.

Thus, like Bissell, Hirschauer does not conform to such simple associations between immobility and inactivity. In an effort to 'absent themselves' from these situations, passengers are active in their inactivity, 'movements, gesture, mimic, sounds—are restrained, the looks are dimmed' (2005: 34). Even though the 'body is motionless, expressionless, soundless' (2005: 34), it appears that these states of passivity and disengagement actually require considerable *practical accomplishment* in order to resolve tensions of 'productivity', 'nervousness', 'restlessness' and 'alertness' (2005: 61).

Further reading

(Vannini 2002; Hirschauer 2005; Bissell 2007; Harrison 2007)

We should remember that these connections and disconnections are incredibly fragile. Turbulence, shocks, bumps and touches can relocate and connect extremely fast. In the context of the car (Katz 1999), even though driving can disconnect and block lots of kinds of relation and intersubjective communications, 'its lack of opportunity for symmetrical interaction may be the key aggravating factor' (Thrift 2004a: 47).

AUGMENTATION

But what sorts of communicational technologies are enabling the mediation of mobilities? What sorts of information communication technologies (ICTs) permit, manage, organize, control or simply alter the movement and sociability of people and things? In this final section of the chapter we can narrow down our focus to several aspects of these technological mobile mediations.

Several authors have bemoaned the distinct lack of research in this area. Graham and Marvin articulate their frustration at how most 'social analyses of cities' continue to 'address urban sociologies, economic development, governance and politics, urban cultures and identities, and urban sociologies and environments', but do so without serious consideration of the 'roles of networked infrastructures in mediating all' (2001: 19). Some of the work to have emerged in this area sits on the apex of a disparate body of research to suggest that the social activity provided by ICTs will destroy the need for physical mobility. So the argument goes that virtual mobilities will negate physical displacement because travelling for face-to-face communication has become unnecessary. It may be put down to the notion that emails are replacing physical letters or because video conferencing will remove the need for the physical conferences.

In the following key ideas box I discuss an influential article in media and communication studies that considers the utility and the potential extension of the concept of mobility for ICT. While the movement of information such as telephone signals and Internet data packets, or the sunk formations of cables, wires and infrastructure, may seem both to be rather dry, technical and irrelevant, the authors argue that they are in fact irrevocably tied to the heart of our concern for mobility with critical social inflections and consequences.

Key idea 5.5: TECHNOLOGICAL MOBILITIES, NETWORKED INFRASTRUCTURES AND INFORMATION COMMUNICATION TECHNOLOGY

In a highly read and cited article Kakihara and Sorensen (Kakihara and Sorensen 2002) set about discussing what import 'mobility' could have for research on ICTs as they worked to expand upon and extend the notion of mobility. Taking mobility as more than a matter of physical displacement, they suggest that it is 'related to the *interaction* they perform — the way in which they [people] interact with each other in their social lives' (2002: 1).

Like many authors before and after them, Kakihara and Sorenson appear frustrated with definitions of mobility that look only at human geographical movement instrumentally. As we have seen so far, this sort of perspective has already received much coverage, yet Kakihara and Sorenson are able to valuably expand the utility of mobility in order to explore the relationship between physical movement and ICTs.

Their article and review looks into how ICTs are vital to the smooth running of modern transport systems, going so far to suggest that they are 'essential "blood vessels" for transportation in a global society' (2002: 1). In their efforts to examine this function, they explore three important dimensions of the dramatic mobilization caused by ICTs. By *spatial* the authors meant the mobilization of both physical objects and virtual spaces such as Web sites. Their second element, *temporality*, referred to the speed and apparent simultaneity ICTs enabled. And by *context*, the authors refer to the 'mobilized situatedness', or rather the circumstances under which this mobilization could be accomplished.

These dimensions provide a useful starting point to explore the conceptual disconnections between 'communication' and 'transportation' McLuhan articulated.

Further reading

(Castells 1996; Kakihara and Sorensen 2002)

From within transportation and communications research, extensive work has unpacked this relationship by posturing what has become known as the concept of mobility *substitution*. These formulations vary in their assessment of the severity of substitution. For some it means the elimination of an entire physical trip. For others, it portends to the foreshortening and alteration of it. Although general consensus on this idea was limited in the 1970s and 1980s (Harkness 1973), by 1998, transportation geographer Susan Hanson argued that 'changes introduced by IT intersect with, and can possibly change, on-the-road processes' (1998: 248). Overall the mediation of mobilities and new virtual mobilities have actually had the effect of stimulating and augmenting real-world travel.

'Complementarity' has been the buzz word for transportation and telecommunications researchers who posit that ICTs and other forms of virtual mobility have augmented and actually created new physical journeys (Salomon 1985, 1986; Mokhtarian 1990, 1991). Pnina Plaut (1997) found that there is clear evidence to suggest how complementarity has been enabled in two distinct ways. First, virtual mobility permits the enhancement of new sorts of mobilities that would have otherwise not occurred. Second, the increased 'efficiency' and productivity of various logistical systems has been enabled, thereby allowing more movements to be performed in smaller amounts of time.

The sustained work of Patricia Mokhtarian (Brown *et al.* 2005; Mokhtarian 2005; Choo *et al.* 2007) has gone further to show how this relation could work both ways. While virtual movement may complement physical mobility, physical mobilities can in turn enhance and lead to even more virtual movements. Transport and telecommunication mobilities increase, it seems, at the same time as each other. As Graham summarizes, 'Overall, transport and telecommunications actually feed off and fuel, more than simply substitute, each other' (Graham 2004a: 254).

Grounds and contexts

The celebration of virtual mobilities and their capacity to remove the need for physical movement, comes paired with the idea that they are somehow emancipated from physical and spatial locations. It is put that virtual mediations allow the unhooking of mobilities and social relations. And yet, we should know by now that these kinds of generalizations of free-flowing and nomadic liberations are never that simple.

Turning again back to the writings of Marvin and Graham (2001), we are presented with the idea that the virtual mediation of mobilities by ICTs are actually dependent upon massively fixed and sunk infrastructural networks, pipelines and *heavy* technological components, composing similar mobility-mooring dynamics presented throughout the book so far. We can surmise a certain *groundedness* to mediated mobilities, which is not to say that mobilities facilitated and mediated by ICTs are necessarily tied down, but that they are able to be so mobile because their infrastructures are fixed. Second, mediated mobilities will often mirror the shape and pattern of their physically placed infrastructural enablers. It is precisely this grounding that Manuel Castells argues has enabled a restructured 'network society' to emerge along the lines and limits of ICTs (see key idea box that follows).

Key idea 5.6: THE NETWORK SOCIETY

For urban sociologist Manuel Castells a new organizational structure of society has emerged that appears to be rooted or organized around the morphology of the network. Although he suggests that network organization had occurred before information technology had provided the 'material basis for its pervasive expansion throughout the entire social structure' (1996: 468), trawling through diverse spheres of social life, Castells argues how networked relations depend upon a networked economy of flow and interaction. Tracing linkages between the 'knowledge information base' of the economy, its global reach and the revolution in Information Technology, these factors have given 'birth to a new and distinctive economic system'. For Castells, the logics and the material lines of such a system have worked to restructure communities and all sorts of social activities.

But underpinning all of these changes appears a spatial material structure, or form of support, which is stratified by three important layers.

1 The material devices or infrastructures that constitute a circuit of electronic impulses in order to construct networks of communication.
2 Certain places act as the organizers or hubs in order to coordinate this flow such as the 'global city' where financial systems and multinational organizations may coordinate. Other points act more as nodes organizing activity in the locality.

3 Actors, or 'dominant, managerial elites' who direct and dominate the interests behind the organization of the networks. Networks are 'enacted, indeed conceived, decided, and implemented by social actors' who regularly traverse very similar trajectories to the networks they organize (Castells 1996: 415).

From work to warfare, Castells shows how networked societies are constituted by quite different social practices and activities, reflecting the ability to move and communicate at speed (almost instantly) and across spatial distances.

Further reading

(Castells 1996; Graham and Marvin 1996, 2001; Dodge and Kitchin 2001)

As Graham has put forward, to look at the traffic of the Internet uncovers the way most of the movement 'actually represents and articulates real places and spaces' (1998; Dodge and Kitchin 2001). Therefore, broadband speed will be dependent upon one's distance away from an exchange and the quality of the line. Wifi access is entirely bound up in the location of Wifi transmitters and the buildings that surround one. In the context of the mobile cellular phone a vast, fixed infrastructure of cables, switches and arrangements need to be in place for talk and communication to occur. For John Agar, 'Mobility, strangely, depends on fixtures' (Agar 2003: 22). That these movements may well support and enable physically mobile activities – from mediated transport systems to tourism – further demonstrates how material mobilities may be constrained by the grounded infrastructures that enable them (Graham 1998: 173). Therefore, the contradiction here is that the barriers to increased mobility can be 'reduced only through the production of particular spaces (railways, highways, airports, teleports, etc.)' (Harvey 1989: 232) and infrastructural fixities. Erik Swyngedouw has similarly argued how communications and transport technologies can really only ever 'liberate activities from their embeddedness in space by producing new territorial configurations', or, in other words, how mobility can only occur through the 'the construction of new (relatively) fixed and confining structures' (Swyngedouw 1993: 306).

Further still, as we think about the grounding of electronic and virtual mobilities that permit and necessitate other physical ones, research has shown how ICTs are embedded within the fabric of the everyday. Mundane infrastructures mediate mundane daily mobilities so much so that societies come to depend upon them. Dodge and Kitchin's research on this issue poses that certain spaces could simply not function without their mediation by computer software and ICTs. These code/spaces are produced through software and informational mobility (Thrift and French 2002) so that a space like an airport becomes a '"a complex set of interlocking assemblages" of ticketing, security, surveillance, flight, traffic control, immigration and many others', that now 'define the practice and experience of air travel itself' (Dodge and Kitchin 2004: 197).

Surveillance

If we establish that mediated mobilities by software and ICTs have become residents in the fabric of the everyday such as the airport terminal, in what other places are these mediations placed and used? To link mediated mobilities with the ever-increasing virtualization of the cyber city suggests that mediated mobilities may have become as ubiquitous as the urban spaces they are tied to (Mitchell 1995; Amin and Thrift 2002; Thrift and French 2002).

In the following passage, I reproduce an extract from Dodge and Kitchin (2005) who detail the possible movements of a fictional character, Elizabeth. Elizabeth lives in London and in the course of just a few hours moving about the city, her movements are constantly mediated and augmented along the way.

> Elizabeth's day starts at 7:00 a.m. After an hour of getting ready, she heads out of the house, turning her iPod on, and walks down Eldon Road, crosses Lordship Lane, and walks along Moselle Street. At the end of the street, she turns right onto The Broadway under the gaze of two private security cameras stationed above an estate agent's. She waits at the curb of Bull Road as three double-decker buses pass. The buses, unbeknown to her, transmit their locations to a small transponder box mounted on a lamppost that updates the estimated arrival time on the "countdown" digital displays along the buses' routes.
>
> (2005: 168)

Before 08:00 a.m., Elizabeth's journey has already encountered and been facilitated by a number of different sorts of mediations. Taking her sound with her, Elizabeth's iPod music allows the listener to enclose themselves in a bubble of personal space (Bull 2007). Her bank's ATM means she can withdraw money from her account instantly. The buses that constitute the milieu of traffic and provide the temporary obstacle to the crossing of the road are being mediated and facilitated by electronic transfer of information.

But there's more to it, let's pick up on the second hour of Elizabeth's journey:

> Just after 8:10, she heads into the Tube station. She waves her "smart" card ticket over a transponder, and the ticket barrier opens, a debit is taken against her card, and she is logged into the Underground monitoring system. Around her, a cluster of five security cameras, part of the Underground's integrated passenger management and security system that covers the entire network, tracks her and the other customers' movements as she descends to the platform where four more cameras are located.
>
> (Dodge and Kitchin 2005: 168)

Now the mediation of Elizabeth's journey has become no less silent but more active, particularly so in the capturing of information. Many of the coded and ICT systems Dodge and Kitchin (2005) refer to might be actually deemed as forms of surveillance. Elizabeth is constantly monitored by closed circuit television (CCTV) cameras along her journey. Her use of her smart pre-payment card on the London Underground enables various management systems to monitor her movements. Without opting in for these mediations she cannot own an Oyster card, while without submitting to the video and electronic monitoring of the ATM machine, she could not withdraw cash. In other words, these systems are entirely necessary to her activities.

What we have revealed are a host of systems aimed at augmenting passenger mobilities by way of and maybe *for* surveillance. In many respects surveillance plays an important role in the augmentation or mediation mobilities in that it may well be necessary for the system to work. In this instance surveillance might be seen as a more benign tool in order to facilitate Elizabeth's movements across the city. In other

Figure 5.3 Oystercard 'pay-as-you-go' travel in London

Source: Copyright © Transport for London 2005

contexts surveillance may well be the purpose of the mediation of mobility itself; mobilities may be mediated *so that* they may be put under control. In ending this section of the chapter, we can briefly dwell on how surveillance (and its enactment by various software and ICT systems) are involved in the mediation of mobilities (see key idea box below).

We have already considered the historical context of workplace surveillance. From Taylor's scientific management of work (see Cresswell earlier) to contemporary monitoring of work email, the surveillance of worker mobilities attempts to assess productivity and efficiency (Marx 1999). In other contexts, the field of tourism research has made recent moves to explore these issues. The subjection of tourists to scrutiny and examination is now an object of considerable public scrutiny and academic analysis in itself. And yet, even though 'tourism mobilities are a vital focus of the new surveillant assemblage', as Morgan and Pritchard write (2005), it is incredibly surprising 'that so little attention has yet been devoted to exploring the powerful discourses and hegemonies which structure this tourism–surveillance dialectic' (Morgan and Pritchard 2005: 125–126).

Key idea 5.7: MOBILITIES AND SURVEILLANCE STUDIES

The emerging area of surveillance studies has sought to investigate all aspects of the relation between practices of surveillance and domination and control. The concept of mobility has become of central importance to this field. As Roger Clarke summarizes, mobility describes the ability for surveillance technologies to move locations and importantly transmit information (Clarke 2003).

There are perhaps two main factors that have raised the subject of mobility to the main priority of the discipline's concern. First, there is a recognition that surveillance is being required to respond to an increasingly more mobile world. Colin Bennett and Priscilla Regan's (2004) editorial in *Surveillance and Society* note how 'people no longer exist and live in fixed locations and spaces . . . within each [social] sphere movement from one activity or place to another, rather than permanence, is likely to be the norm.' In other words, it is because of all the moving, transporting, migrating and travelling that surveillance must occur. It is because we, and increasingly, things, are so mobile that they must be put under surveillance.

Second, it is in the response of surveillance to the requirements of mobilities that surveillance practices have become mobile in themselves. David Lyon writes how it is 'important to remember that the growth of global surveillance is not a conspiracy. The primary reason why surveillance is globalizing is that mobility is a fundamental feature . . . that now dominates the world' (Lyon 2007: 121).

Further reading

(Lyon 2003a, 2007; Adey 2004)
See the important Web site, http://www.surveillance-and-society.org

Change appears to be driving the securitization of everyday mobilities, enabling the increasing fluidity of surveillance; thus, Lyon writes, 'today surveillance itself is part of the flow' (2003b). The increasing mobility of surveillance mediation, it seems, is enabled by the increasing informationalization of the world and the ability to move this information around. What has been called dataveillance (Clarke 1988) refers to the systematic way that data about ourselves – the traces of our movements and behaviours – may be collected and monitored. Lyon

(2002) describes the actual trend of capturing and creating this data as the process of the 'phenetic fix'. A process 'endemic' to modern surveillance systems, the phenetic fix turns physical and practical behaviours into data that can be managed and used.

Thus, lots of interesting and important work has demonstrated how physical mobilities may be tracked according to the digital signature they leave. These processes enable Dodge and Kitchin's commuter in London to be followed by her electronic and digital transactions. A key facet in these debates is that information moves and intersects with real physical journeys of people. Marieke de Goede and Louise Amoore's (2008) recent overview of these processes highlights the use of travel data in airline security, surveillance and the monitoring of financial transactions. In examining the Passenger Name Record (PNR) agreement of 2004, between European Airlines and the United States, they point out how airlines were forced to transmit some 34 items of information on passenger behaviour and characteristics. The data was shuttled back and forth through an electronic transaction between European Airlines and the U.S. Department for Homeland Security (Bennett 2006; see also Bigo 2006). The authors summarize how the transmissions or the *transactions* are 'represented as a means of reconciling mobility with security or, as for the Department of Homeland Security, a way to "keep the data flowing and the planes flying" (Amoore and de Goede 2008: 1).

Surveillance is thus increasingly mobile because data about mobile individuals may be easily moved about between databases and the organizations that hold them. Given that we live in a 'personal information economy' everyday mobilities may pen their own digital signatures embedded within objects such as passports and visas, which are passed from place to place the world over' (Lyon 2007: 121). This is occurring by the fact that data are embedded in moving objects and because the repositories where this information may be kept are becoming much more porous and open. Kevin Haggerty and Richard Ericson's seminal article on what they call the surveillant assemblage outlines how surveillance systems are that bit more tied together: '[T]he desire to bring systems together, to combine practices and technologies and integrate them into a larger whole' (2000: 610) appears to be leading the increasing connectivity and amount of data flow. Information sharing is something we might all experience just by picking up our post, wondering why we are receiving just so much junk mail. Our personal information may have an economic value; it is gathered up and maybe sold off to marketers trying to advertise the latest financial product. Or as Amoore and De Goede (2008) showed us earlier, it may be valued in terms of security, so that organizations may be compelled to share what they know about us and our movements.

But what does all of this really mean? Many of you might say to yourselves, 'Well, I've done nothing wrong, so does it really matter?' Arguing that the implications go further than the invasion of privacy, it is put forward that it doesn't matter whether you have done anything wrong or not. Lyon (2003b) is perfectly serious when he claims that the old adage that 'if you have nothing to hide, you have nothing to fear' is no longer very true. For Lyon and many others, mobile surveillance has begun to systematically affect people's mobilities and life chances because it works as an active system of discrimination. Some authors have surmised that the fluidic 'mobile publics' of modern societies are actually 'publics that are often prioritized, enacted and kept apart by hidden worlds of software-sorting' (Graham 2005: 564). And examples can be seen in various aspects of everyday life, from the prioritization of preferred customers in telephone call centre queuing systems – the mundane spheres of even the telephone call centre – may hold just as evident inequalities (2005). Elsewhere David Holmes (2004) shows how a road-tolling programme in Sydney prioritizes wealthy car owners.

Mobile method 5.2: TRACING MOBILITIES

The ability to monitor mobilities, as shown above, provides innovative new ways to capture and record mobility and often in real time. Global positioning satellite systems (GPS) mean it is possible to track the position of willing respondents (Parks 2005) at a distance quickly and easily.

Regular technologies useful to qualitative forms of research method are becoming – as we have seen – more mobile in ways that open out new possibilities. 'Equipment that was once the preserve of secret agents is now within the reach of most academic departments' write Evans *et al.* (2008: 1,272) in their recent review. Helmet-mounted cameras utilized by cycling researcher Justin Spinney means the researcher can record a much closer perspective on an ethnographic encounter and give its viewer a first-person perspective of moving in an environment. As discussed earlier in the chapter, Laurier's development of video methods allows the complex practices of mobile work to be demonstrated.

Other bio-centred biomedical monitoring devices may give further detail to the body-in-motion. Heart rate monitors and body temperature devices assess 'the bodily affects and engagements with different environments' (Hein *et al.* 2008) while pedometers trace quantitative distance. Also don't forget that many new recording devices are digital. Therefore, much of this information has become remarkably mobile too, given the ability to transmit this data easily across the Internet and to share it and make comparisons with other studies and data sets.

All of this mobility and digitalization has meant the increase in the traces of mobility. Data entries and electronic records are a recent addition to this. Trevor Paglen's (2006) fascinating work on 'extraordinary rendition' shows how the movements of terror flights conducted by the CIA have left behind long data trails of records.

In practice

- Mobile and lightweight equipment enables the recording of more experiential research activities and at a distance
- Databases and systems of records may allow the tracing of past and present movements
- Consideration should be given to ethical and privacy issues

Further reading

(Laurier 2004; Spinney 2006; Hein *et al.* 2008)

Mobile prostheses

The discussed mediating technologies initiate the speeding up, slowing down, augmentation and capturing of mobilities. While most are grounded in some way, these and other technologies successfully enable the smooth liberation of people from places, permitting social practices to become both mobile and fluid. As mentioned during

Elizabeth's walk (above), mobile prosthesis such as her iPod enabled her to bring her previously and relatively static music collection with her on her journey (bringing potentially all sorts of cultural worlds with her, Bull 2001). In this final section of the chapter, we can consider how mobile bodily technologies such as the mobile phone have made social actions and practices more mobile.

At the beginning of the chapter we thought through the sorts of new work practices that are being conducted in places like the car or the aeroplane, allowing 'travel time' to be seen in much more productive ways. Travel time becomes not lost 'dead' time, but it may be turned into something much more usable and workable (Lyons and Urry 2005; Jain and Lyons 2008). John Agar (2003) uses an advert from British Telecom in 1986 to demonstrate how dead time could be resurrected through the purchase of their first mobile phones.

TURNING IDLE TIME INTO PRODUCTIVE TIME

> When you're away from your office and your phone, you're effectively out of touch with your business. You can't be contacted. Nor can you easily make contact yourself. Take a mobile telephone – a Cell phone – with you and you get a double benefit. You're totally in touch, ready to take instant advantage of business opportunities when and where they occur. And you can make maximum effective use of 'dead time' – time spent travelling – turning it into genuine productive hours.
>
> (from Agar 2003: 83)

What we have through the mobile phone is a *mobile-ization* which enables and 'reinforces mobile lifestyles and physically dispersed relationships' (Wellman 2001: 239). Fixed and sunk infrastructural networks provide the mooring to the mobile phone that can be carried about across spaces, liberating its user from 'place and group'. The point is, however, that the liberation is not complete. The phone permits a 'constant touch' or a 'perpetual contact' to be maintained to people, actions and practices (Katz and Aakhus 2002). In this way it is not the moorings of social relations which are lost, just those of space (within reason as phone signals depend upon nearby transmitters).

The mobile phone enables the mobility of its user to interact and connect on the move, to work, to communicate and to socialize. Bauman (2003) expands on these thoughts as he considers how the mobile phone, in all this liquidity, can maintain associations and connections. Bauman writes,

> You *stay connected* – even though you are constantly on the move, and though the invisible senders and recipients of calls and messages move as well, all following their own trajectories. Mobiles are for people on the move.
>
> (2003: 59)

Rendering an image of something like a network diagram, Bauman describes how the nodes on the network could be considered one's friends or work contacts. Bauman then sets this diagram in motion; one's friends, work colleagues or even distant relatives are always transient throughout their own daily life courses, routines, journeys and vacations. But the mobile phone allows the lines between these networks to hold firm even as they stretch, skew and overlap one another. The associations and connections are the stable point in 'the universe of moving objects'. The connections remain even though geographically the phone's user may have moved, for 'Connections are rocks among the quicksand. On them you can count' (Bauman 2003: 59).

Connecting successfully to the distant and elsewhere can mean disconnection with the local and the contextual. Touching upon a manner of withdrawal we discussed earlier, Barry Wellman recounts an experience on Toronto's streetcar as he and his wife listened in, to a young women who embarked on an 'intensely romantic' conversation with her lover. 'She seemed oblivious to my sitting next to her,' Wellman explains, 'Her intense involvement in her private conversation – and her loud voice intruding on our soft conversation – appropriated public space for her own needs' (Wellman 2001: 239–240). In this light, a mobile phone user – while on the move – appears able to carve out a private space for themselves.

Kopomma (2004) describes this as private 'bubble' is inflated outwards as the user 'withdraws from the social situation'. Like the active inaction discussed earlier, the 'speaker's behavior is characterized by absence and a certain introversion. Staring into space as well as smiling are both indications of withdrawal into mobile phone sociability' (Kopomma 2004: 270). Ringing up a loved one can make this bubble feel protective during the passage home through a threatening neighbourhood. The consistent checking of one's phone, once alerted by a vibration or a noise, and responding with a tapping of buttons is a familiar set of practices associated with the answering of an SMS message (Licoppe 2004; Licoppe and Inada 2006).

These inwardly private expressions can move outwards as Wellman explains. Strangers can be drawn into the bubble of a loud phone conversation as their concentration is broken, their conversation interrupted or they are awakened by a loud text-message alert. With the advent of in-built facilities in a mobile phone, such as music players, video recorders – cameras too – the mobile phone usage has taken on multiple dimensions: while on the move music can be listened to and swapped, and friends taking pictures can transfer them instantly to another's phone. I recently sat on the bus and couldn't help but notice a young man on the seat in front of me using a video phone to speak to a friend. Seeing the friend on the phone's display I was immediately interested by this clever piece of technology, before I quickly averted my gaze as I felt literally drawn into their private conversation (if I could see the friend they could presumably have seen me peering from behind the seat!). These personal bubbles can overlap and interfere with one another in surprising and unpredictable ways.

CONCLUSION

What does it mean to mediate? In the context of our study of mobility the answer is manifold; we have seen several different kinds of mediation. Mobilities seem to always carry something inside; no matter how many times you remove a layer, one after another, something has stown away. Mobilities in other words are parasitic. Piggybacking other mobilities, mobile bodies are facilitated in their movement by transportation technologies as prevalent as the car, the train, the bus and others. These mediating technologies mediate many forms of mobility; indeed, almost every mobility is mediated by something.

Some of these mediated mobilities require vast, fixed infrastructures in order to facilitate their journey and other systems of mediation in order to control and regulate them. Surveillance was shown as one important mechanism. For importantly, mediation plays a key role in the enaction of mobilities we considered in the previous chapter.

Practices of surveillance and security may socially sort mobilities into favoured high-priority groups and less favoured low-priority groups, affecting people's life chances and future mobilities – their motility.

The mobile phone as we have just seen can liberate social relations and maintain constant communication. On the other hand, the walking boot may get between by deadening communication from body to ground.

Connection and disconnection were thus discussed as a primary complication of mediated mobilities. Mediation can serve as a buffering actor between relations or it can conduit improved communications and closer connections.

6

CONCLUSION

To conclude many of the themes and connections of this book, I want to end in a rather unorthodox way – through a discussion of Audrey Niffenegger's novel *The Time Traveller's Wife* – which tells a complex story of two people attempting to connect: Clare and Henry. Their efforts to be together are constantly frustrated by Henry's chrono-displacement, a genetic abnormality that endows him with the ability to be unwillingly transported through time and space.

We could take Henry and Clare's struggle to meet up – to synchronize their selves at different points in their lives – as some sort of allegory of contemporary mobilities and the shape of mobile existence. The desire for synchronicity – *proximity* – has become a signature of many of the mobile worlds we have seen. Running on different timetables people struggle to match up with each other and the schedules of the mobile infrastructual and transportation systems they rely upon to get them from one place to another. Henry's disorientating experiences of arrival could say something many of us would identify with as we step out of an aircraft to the chaos of an airport terminal, or a migrant's experience of an adventure into a new city. Henry's movements are tainted with the feelings of loss, regret and longing we have seen in stretched-out social relations of migrant belongings. On the other hand, his movements are full of happiness at a welcome arrival and finding himself back in the security of his home and the sanctity of place.

The uncertainty of his slippages resemble contemporary assertions of threatening circulations as well as unruly and unmanageable fluidities. All the while, his capacity to spring into spaces where he shouldn't be, and his tendency to transgress the normative codes of places where he shouldn't be, send him thrashing headfirst against practices of regulation, surveillance, rule and the law. As he struggles to adapt to his uncertain displacements, Henry becomes an expert at trespassing, minor thievery and border crossing; he is less good at avoiding motion sensors.

At the same time, perhaps, the metaphor doesn't quite work. Henry's journey through space and time appears to negate some of the strongest arguments I have been making about mobility. Henry is literally a monad. In fact his mobility is not really mobility in the way we have thought about it. It involves an instantaneous slippage or a sliding between the layers of geography and history without effort, experience or knowledge. Henry is well and truly alone. He travels without anyone. He travels even without things – an absence of clothes is a troublesome worry. These movements could stand for the feelings of isolation a regular traveller might undergo, but according to the many different ideas about mobility we've taken, Henry's mobility doesn't exist in practice. Henry's movements are movements *without* – to the extreme – and they sharply contrast with the variety of ways that mobility appears to always occur *with* – with someone or something else. Whether these are people or things that one is indifferent to or close to, whether they mediate one's journey, or whether they are things and others one brings with one – friends, family, carers, business colleagues, baggage, shopping, laptop, sustenance – mobility is never ever really singular. Far from an instantaneous journey of departure and arrival, mobility – like a train journey – is experienced. It might be quick, or slow; it might be easily described or thoroughly non-representational, but it is still something done.

So far so good, but considering the story for a bit longer, perhaps such a conclusion is a bit too convenient. So we might lose a sense of Henry-being-mobile – his experiences, sensations and relations-with. But on the other hand, the book is called *The Time Traveller's Wife*, and we are told not only of Henry but also Clare, Henry's wife. Let us quickly not forget that Clare's subservience to Henry's arrivals and departures might conform to some of the gendered inequities of mobility we have seen. But more to the point we learn of impact; we learn of those geometries of effect. We learn about mobility in-relation-to. We learn of how Henry, Clare and their friends are all placed in very different ways to Henry's movements. We learn of belonging to someone else; the importance of places and people where one might fit in or call home. And we learn of the wrench of leaving – the feeling of a tie broken.

Moreover, while we have little idea of the experience of travel, the effects it has on Henry and Clare are clear to see. As we have discussed mobility requires effort, it demands work and it costs lives. Henry's slippages are sometimes painful, but mostly they are exhausting and they often make him nauseous. Sometimes it is smoother, easy and he is joyful on turning up, although he may feel dread in the anticipation of a departure. Indeed, even though the passivity of Henry's mobility seems initially so unlikely, it speaks to the experiences of those who are not so in control, those who are forced to move when they don't want to – the experiences of mobility that are much more inactive, submissive or acquiescent.

This book has argued how the concept 'mobility' courses through the key issues, problems and developments of our social world, as well as made efforts to understand and make sense of them. From disparate disciplines, ideas, case studies and techniques, we need to see how both mobilities and our ideas, methods and findings run alongside one another. Moving on mobilities research demands that we continue to do and appreciate this moving-with.

BIBLIOGRAPHY

Aas, K. F. (2007) 'Analysing a world in motion: global flows meet criminology of the other', *Theoretical Criminology*, 11: 283–303.

Abler, R., Adams, J. S. and Gould, P. (1971) *Spatial organization: the geographer's view of the world*, Englewood Cliffs, NJ; Hemel Hempstead: Prentice-Hall.

Adams, P. C. (2001) 'Peripatetic imagery and peripatetic sense of place', in Adams, P. C., Hoelscher, S. and Till, K. (eds.) *Textures of place: exploring humanist geographies*, Minneapolis: University of Minnesota Press.

Adams, R. M. (1974) 'Anthropological perspectives on ancient trade', *Current Anthropology*, 15: 239–249.

Adey, P. (2004) 'Secured and sorted mobilities: examples from the airport', *Surveillance and Society*, 1: 500–519.

——— (2006a) 'Airports and air-mindedness: spacing, timing and using Liverpool Airport 1929–39', *Social and Cultural Geography*, 7: 343–363.

——— (2006b) 'If mobility is everything then it is nothing: towards a relational politics of (im)mobilities', *Mobilities*, 1: 75–94.

——— (2007) "May I have your attention': airport geographies of spectatorship, position and (im)mobility', *Environment and Planning D – Society and Space*, 3: 515–536.

——— (2008) 'Airports, mobility, and the calculative architecture of affective control', *Geoforum*, 39: 438–451.

Adey, P., Budd, L. and Hubbard, P. (2007) 'Flying lessons: exploring the social and cultural geographies of global air travel', *Progress in Human Geography*, 31: 773–791.

Agamben, G. (2005) *State of exception*, Chicago, Il; London: University of Chicago Press.

Agar, J. (2003) *Constant touch: a global history of the mobile phone*, Cambridge: Icon.

Ahmed, S. (2004) *The cultural politics of emotion*, Edinburgh: Edinburgh University Press.

Ali, S. H., and Keil, R. (2008) *Networked disease: emerging infections in the global city*, Malden, Mass.: Blackwell.

Allen, J. and Hamnet, C. (1995) *A shrinking world?*, Oxford: Open University Press.

Alliez, E. (2004) *The signature of the world: or, what is Deleuze and Guattari's philosophy?*, New York; London: Continuum.

Amin, A. and Thrift, N. (2002) *Cities: reimagining the urban*, Cambridge: Polity.
Amoore, L. and De Goede, M. (2008) 'Transactions after 9/11: the banal face of the preemptive strike', *Transactions – Institute of British Geographers*, 33: 173–185.
Anderson, B. (2006) 'Becoming and being hopeful: towards a theory of affect', *Environment and Planning D*, 24: 733–752.
Anderson, J. (2004) 'Talking whilst walking: a geographical archaeology of knowledge', *Area – Institute of British Geographers*, 36: 254–261.
Angus, J., Kontos, P., Dyck, I., McKeever, P. and Poland, B. (2005) 'The personal significance of home: *habitus* and the experience of receiving long-term home care', *Sociology of Health and Illness*, 27: 161–187.
Appadurai, A. (1990) 'Disjuncture and difference in the global cultural economy', in Featherstone, M. (ed.) *Global culture: nationalism, globalization and world culture*, London: Sage.
Appadurai, A. (1995) 'The production of locality', in Fardon, R. (ed.) *Counterworks: managing the diversity of knowledge*, London: Routledge.
Atkinson, D. (1999) 'Nomadic strategies and colonial governance: domination and resistance in Cyrenaica, 1923–1932', in Sharp, J. (ed.) *Geographies of domination/resistance; entanglements of power*, London: Routledge.
Augé, M. (1995) *Non-places: introduction to an anthropology of modernity*, New York: Verso.
Auster, P. (1987) *The New York trilogy*, London: Faber.
Bachelard, G. (1988) *Air and dreams: an essay on the imagination of movement*, Dallas: Dallas Institute Publications, Dallas Institute of Humanities and Culture.
Bahnisch, M. (2000) 'Embodied work, divided labour: subjectivity and the scientific management of the body in Frederick W. Taylor's 1907 "Lecture on Management"', *Body and Society*, 6: 51–68.
Bajc, V. (2007) 'Creating ritual through narrative, place and performance in evangelical Protestant pilgrimage in the Holy Land', *Mobilities*, 2: 395–412.
Bajc, V., Coleman, S. and Eade, J. (2007) 'Introduction: mobility and centring in pilgrimage', *Mobilities*, 2: 321–329.
Bale, J. (1996) *Kenyan running: movement culture, geography, and global change*, London; Portland, Oreg.: Frank Cass.
——— (2004) *Running cultures: racing in time and space*, London; New York: Routledge.
Barber, L. G. (2002) *Marching on Washington: the forging of an American political tradition*, Berkeley: University of California Press.
Barnes, T. J. (2008) 'Geography's underworld: the military–industrial complex, mathematical modelling and the quantitative revolution', *Geoforum*, 39: 3–16.
Barnes, T. J. and Farish, M. (2006) 'Between regions: science, militarism, and American geography from World War to Cold War', *Annals of the Association of American Geographers*, 96: 807–826.
Bartling, H. (2006) 'Suburbia, mobility, and urban calamities', *Space and Culture*, 9: 60–62.
Bassett, K. (2004) 'Walking as an aesthetic practice and a critical tool: some psychogeographic experiments', *Journal of Geography in Higher Education*, 28: 397–410.
Baudrillard, J. (1988) *America*, London: Verso.
Bauman, Z. (1998) *Globalization*, Cambridge: Polity Press.
——— (2000) *Liquid modernity*, Cambridge; Malden, Mass.: Polity Press; Blackwell.

——— (2003) *Liquid love: on the frailty of human bonds*, Cambridge: Polity Press; Malden, Mass.: Distributed in the USA by Blackwell.

Bechmann, J. (2004) 'Ambivalent spaces of restlessness: ordering (im)mobilities at airports', in Bærenholdt, J. O. and Simonsen, K. (eds.) *Space odysseys: spatiality and social relations in the 21st century*, Aldershot: Ashgate.

Beck, U. (2000) *What is globalization?*, Malden, Mass.: Polity Press.

——— (2006) *The cosmopolitan vision*, Cambridge: Polity.

Beckmann, J. (2001) 'Automobility – a social problem and theoretical concept', *Environment and planning D*, 19: 593–608.

Benjamin, W. (1973) *Charles Baudelaire: a lyric poet in the era of high capitalism*, London: NLB.

——— (1985) *One-way street and other writings*, London: Verso.

——— (1986) *Illuminations*, New York: Schocken Books.

——— (1999) *The Arcades Project*, Cambridge, Mass.: Belknap Press of Harvard University Press.

Bennett, C. (2006) 'What happens when you book an airline ticket (revisited): the computer assisted passenger profiling system and the globalization of personal data', in Zureik, E. and Salter, M. B. (eds.) *Global surveillance and policing: borders, security, identity*, Cullompton: Willan.

Bennett, C. and Regan, P. (2004) 'Surveillance and mobilities', *Surveillance and Society*, 1: 449–455.

Bergson, H. (1911) *Creative evolution*, New York: H. Holt and Company.

——— (1950) *Matter and memory*, London, New York: G. Allen and Unwin; Macmillan.

Berman, M. (1983) *All that is solid melts into air: the experience of modernity*, New York: Verso.

Bhabha, H. K. (1994) *The location of culture*, London: Routledge.

Bigo, D. (2006) 'Security, exception, ban and surveillance', in Lyon, D. (ed.) *Theorizing surveillance: the panopticon and beyond*, Culhompton: Willan.

Bissell, D. (2007) 'Animating suspension: waiting for mobilities', *Mobilities*, 2: 277–298.

——— (2008) 'Comfortable bodies: sedentary affects', *Environment and Planning A*, 40: 1697–1712.

Blomley, N. K. (1992) 'The business of mobility – geography, liberalism, and the Charter of Rights', *Canadian Geographer – Geographe Canadien*, 36: 236–253.

——— (1994a) *Law, space, and the geographies of power*, New York; London: Guilford.

——— (1994b) 'Mobility, empowerment and the rights revolution', *Political Geography*, 13: 407–422.

Blunt, A. (1994) *Travel, gender and imperialism: Mary Kingsley and west Africa*, New York: Guilford.

——— (2005) *Domicile and diaspora: Anglo-Indian women and the spatial politics of home*, Oxford: Blackwell.

——— (2007) 'Cultural geographies of migration: mobility, transnationality and diaspora', *Progress in Human Geography*, 31: 684–694.

Blunt, A. and Dowling, R. M. (2006) *Home*, London: Routledge.

Bohm, S., Jones, C., Land, C. and Paterson, M. (2006) 'Introduction: impossibilities of automobility', *Sociological Review*, 54: 1–16.

Bonsall, P. and Kelly, C. (2005) 'Road user charging and social exclusion: the impact of congestion charges on at-risk groups', *Transport Policy*, 12: 406–418.

Borden, I. (2001) *Skateboarding, space and the city: architecture and the body*, Oxford; New York: Berg.

Bourdieu, P. (1977) *Outline of a theory of practice*, Cambridge: Cambridge University Press.

——— (1984) *Distinction: a social critique of the judgement of taste*, London: Routledge and Kegan Paul.

Bourke, J. (2005) *Fear: a cultural history*, London: Virago Press.

Bowlby, R. (2001) *Carried away: the invention of modern shopping*, New York: Columbia University Press.

Braidotti, R. (1994) *Nomadic subjects: embodiment and sexual difference in contemporary feminist theory*, New York: Columbia University Press.

Brand, S. (1994) *How buildings learn: what happens after they're built*, New York; London: Viking.

Braun, B. (2007) 'Biopolitics and the molecularization of life', *Cultural Geographies*, 14: 6–28.

Brennan, T. (2003) *The transmission of affect*, Ithaca, NY; London: Cornell University Press.

Brenner, N. (1998) 'Between fixity and motion: accumulation, territorial organization and the historical geography of spatial scales', *Environment and Planning D – Society and Space*, 16: 459–481.

Brown, C., Balepur, P. and Mokhtarian, P. L. (2005) 'Communication chains: a methodology for assessing the effects of the Internet on communication and travel', *Journal of Urban Technology*, 12: 71–98.

Brown, L. A. (1968) *Diffusion dynamics: a review and revision of the quantitative theory of the spatial diffusion of innovation*, Lund: Gleerup.

——— (1981) *Innovation diffusion: a new perspective*, London: Methuen.

Brown, L. A., Malecki, E. J., Gross, S. R., Shrestha, M. N. and Semple, R. K. (1974) 'Diffusion of cable television in Ohio – case study of diffusion agency location patterns and processes of polynuclear type', *Economic Geography*, 50: 285–299.

Bruno, G. (1992) 'Streetwalking around Plato's Cave', *October*, 60: 110–129.

——— (2002) *Atlas of emotion: journeys in art, architecture, and film*, New York: Verso.

Buck-Morss, S. (1989) *The dialectics of seeing: Walter Benjamin and the Arcades project*, Cambridge, Mass.: MIT Press.

Bull, M. (2001) 'The world according to sound: investigating the world of Walkman users', *New Media and Society*, 3: 179–198.

——— (2007) *Sound moves: iPod culture and urban experience*, London: Routledge.

Buttimer, A. and Seamon, D. (1980) *The human experience of space and place*, New York: St. Martin's Press.

Calhoun, C. (2002) 'The class consciousness of frequent travelers: toward a critique of actually existing cosmopolitanism', *South Atlantic Quarterly*, 101: 869–898.

Canetti, E. (1962) *Crowds and power*, New York: Viking Press.

Canzler, W., Kaufmann, V. and Kesselring, S. (eds.) (2008) *Tracing mobilities: towards a cosmopolitan perspective*, Aldershot, Ashgate.

Castells, M. (1996) *The rise of the network society*, Oxford: Blackwell.

——— (1997) *The power of identity*, Malden, Mass.; Oxford: Blackwell.

——— (2000) *End of millennium*, Oxford: Blackwell.

Castree, N. (2005) *Nature*, London: Routledge.

Cavanaugh, W. T. (2008) 'Migrant, tourist, pilgrim, monk: mobility and identity in a global age', *Theological Studies*, 69: 340–356.

Certeau, M. D. (1984) *The practice of everyday life*, Berkeley: University of California Press.

Chambers, I. (1986) *Popular culture: the metropolitan experience*, London; New York: Methuen.

Chang, S. E. (2004) 'Transportation geography: the influence of Walter Isard and regional science', *Journal of Geographical Systems*, 6: 55–69.

Chatty, D. and Colchester, M. (2002) *Conservation and mobile indigenous peoples: displacement, forced settlement, and conservation*, New York; Oxford: Berghahn Books.

Choo, S., Lee, T.Y. and Mokhtarian, P. L. (2007) 'Do transportation and communications tend to be substitutes, complements, or neither? US Consumer Expenditures Perspective, 1984–2002', *Transportation Research Record*, 2010: 123–132.

Christaller, W. (1966) *Central places in Southern Germany*, London: Prentice-Hall.

Chyba, C. F. (2002) 'Toward biological security', *Foreign Affairs*, 81: 122–137.

Clarke, R. (1988) 'Information technology and dataveillance', *Communication ACM*, 31: 498–512.

——— (2003) Wireless transmission and mobile technologies: http://www.anu.edu.au/people/Roger.Clarke/EC/WMT.html

Cliff, A. D., Haggett, P., Ord, J. K. and Versey, G. R. (1981) *Spatial diffusion: an historical geography of epidemics in an island community*, Cambridge: Cambridge University Press.

Cloke, P., Goodwin, M., Milbourne, P. and Thomas, C. (1995) 'Deprivation, poverty and marginalization in rural lifestyles in England and Wales', *Journal of Rural Studies*, 11: 351–366.

Cloke, P. J., Goodwin, M. and Milbourne, P. (1997) *Rural Wales: community and marginalization*, Cardiff: University of Wales Press.

Cohen, S. (1995) 'Sounding out the city: music and the sensuous reproduction of place', *Transactions – Institute of British Geographers*, 20: 434–446.

——— (2005) 'Country at the heart of the city: music, heritage, and regeneration in Liverpool', *Ethnomusicology*, 49: 25–48.

Comaroff, J. and Comaroff, J. (2002) 'Alien-nation: zombies, immigrants, and millennial capitalism', *South Atlantic Quarterly*, 101: 779–806.

Connell, J. and Gibson, C. (2003) *Sound tracks: popular music, identity and place*, London: Routledge.

Conradson, D. and Latham, A. (2005) 'Transnational urbanism: attending to everyday practices and mobilities', *Journal of Ethnic and Migration Studies*, 31: 227–234.

——— (2007) 'The affective possibilities of London: antipodean transnationals and the overseas experience', *Mobilities*, 2: 231–254.

Conradson, D. and McKay, D. (2007) 'Translocal subjectivities: mobility, connection, emotion', *Mobilities*, 2: 167–174.

Cook, I. (2004) 'Follow the thing: Papaya', *Antipode*, 36: 642–664.

——— (2006) 'Geographies of food: following', *Progress in Human Geography*, 30: 655–666.

Cook, I., Crang, P. and Thorpe, M. (1998) 'Biographies and geographies: consumer understandings of the origins of foods', *British Food Journal*, 100: 162–167.

Cooper, M. (2006) 'Pre-empting emergence: the biological turn in the War on Terror', *Theory Culture and Society*, 23: 113–136.

Coward, M. (2004) 'Urbicide in Bosnia', in Graham, S. (ed.) *Cities, war and terrorism*, Oxford: Blackwell.

——— (2006) 'Against anthropocentrism: the destruction of the built environment as a distinct form of political violence', *Review of International Studies*, 32: 419–437.

Crang, M. (2001) 'Rhythms of the city: temporalised space and motion' in May, J. and Thrift, N. J. (eds.) *Timespace: geographies of temporality*, London: Routledge.

——— (2002a) 'Between places: producing hubs, flows, and networks', *Environment and Planning A*, 34: 569–574.

——— (2002b) 'Rethinking the observer: film, mobility and the construction of the subject', in Cresswell, T. and Dixon, D. (eds.) *Engaging film: geographies of mobility and identity*, London: Rowman and Littlefield.

Crary, J. (1999) *Suspensions of perception: attention, spectacle, and modern culture*, Cambridge, Mass.: MIT Press.

Crawford, M. (1994) 'The world in a shopping mall', in Sorkin, M. (ed.) *Variations on a theme park: the new American city and the end of public space*, New York: Hill and Wang.

Cresswell, T. (1993) 'Mobility as resistance – a geographical reading of kerouac on the road', *Transactions of the Institute of British Geographers*, 18: 249–262.

——— (1996) *In place/out of place: geography, ideology, and transgression*, Minneapolis; London: University of Minnesota Press.

——— (1997) 'Imagining the nomad: mobility and the postmodern primitive', in Benko, G. and Strohmayer, U. (eds.) *Space and social theory: interpreting modernity and post-modernity*. Oxford, England; Cambridge, Mass.: Blackwell.

——— (1999a) 'Embodiment, power and the politics of mobility: the case of female tramps and hobos', *Transactions – Institute of British Geographers*, 24: 175–192.

——— (1999b) 'Falling down: resistance as diagnostic', in Sharp, J. (ed.) *Geographies of domination/resistance; entanglements of power*. London: Routledge.

——— (2001) 'The production of mobilities', *New Formations*, 43: 11–25.

——— (2002) 'Guest editorial Bourdieu's geographies: in memorium', *Environment and Planning D*, 20: 379–382.

——— (2004) *Place: a short introduction*, Malden, Mass.: Blackwell.

——— (2006a) *On the move: the politics of mobility in the modern west*, London: Routledge.

——— (2006b) 'The right to mobility: the production of mobility in the courtroom', *Antipode*, 38: 735–754.

Cresswell, T. and Dixon, D. (2002) *Engaging film: geographies of mobility and identity*, Lanham, Md.: Rowman and Littlefield.

Cromley, E. K. and McLafferty, S. L. (2002) *GIS and public health*, New York; London: Guilford.

Cronin, A. (2008) 'Mobility and market research: outdoor advertising and the commercial ontology of the city', *Mobilities*, 3: 95–115.

Cronin, A. M. (2006) 'Advertising and the metabolism of the city: urban space, commodity rhythms', *Environment and Planning D*, 24: 615–632.

Cunningham, H. and Heyman, J. M. C. (2004) 'Introduction: mobilities and enclosures at borders', *Identities*, 11: 289–302.

Cwerner, S. B. (2006) 'Vertical flight and urban mobilities: the promise and reality of helicopter travel', *Mobilities*, 1: 191–215.

D'Souza, A. and McDonough, T. (2006) *The invisible flâneuse?: gender, public space and visual culture in nineteenth-century Paris*, Manchester: Manchester University Press.

Damasio, A. R. (2000) *The feeling of what happens: body and emotion in the making of consciousness*, London: W. Heinemann.

Dant, T. (2004) 'The driver-car', *Theory Culture and Society*, 21: 61–80.
Davis, M. (1990) *City of quartz: excavating the future in Los Angeles*, London: Vintage, 1992.
De Landa, M. (1997) *A thousand years of nonlinear history*, New York: Zone Books.
Debord, G. (1970) *Society of the spectacle*, Detroit: Black and Red.
Deleuze, G. (1988) *Spinoza, practical philosophy*, San Francisco: City Lights Books.
Deleuze, G. and Guattari, F. (1988) *A thousand plateaus: capitalism and schizophrenia*, London: Athlone Press.
DeParle, J. (2007) Fearful of restive foreign labor, Dubai eyes reforms, *New York Times*: http://www.nytimes.com/2007/08/06/world/middleeast/06dubai.html
Der Derian, J. (1990) 'The (s)pace of international relations: simulation, surveillance, and speed', *International Studies Quarterly*, 34: 295–310.
Derudder, B., Devriendt, L. and Witlox, F. (2007) 'Flying where you don't want to go: an empirical analysis of hubs in the global airline network', *Tijdschrift Voor Economische En Sociale Geografie*, 98: 307–324.
Derudder, B., Witlox, F., Faulconbridge, J. and Beaverstock, J. (2008) 'Airline data for global city network research: reviewing and refining existing approaches', *Geojournal*, 71: 5–18.
Dewsbury, J. D. (2000) 'Performativity and the event: enacting a philosophy of difference', *Environment and Planning D-Society and Space*, 18: 473–496.
——— (2003) 'Witnessing space: "knowledge without contemplation"', *Environment and Planning A*, 35: 1907–1932.
Dhagamwar, V., De, S. and Verma, N. (2003) *Industrial development and displacement: the people of Korba*, New Delhi; London: Sage Publications.
Dillon, M. (2007) 'Governing terror: the state of emergency of biopolitical governance', *International Political Sociology*, 1: 7–28.
Dillon, M. and Lobo-Guerrero, L. (2008) 'Biopolitics of security in the 21st century', *The Review of International Studies*, 34: 265–292.
Dimendberg, E. (1995) 'The will to motorization, cinema, highways, and modernity', *October* 73: 90–137.
Dodge, M. and Kitchin, R. (2001) *Atlas of cyberspace*, Harlow, England: Addison-Wesley.
——— (2004) 'Flying through code/space: the real virtuality of air travel', *Environment and planning A*, 36: 195–211.
——— (2005) 'Code and the transduction of space', *Annals – Association of American Geographers*, 95: 162–180.
——— (2007) 'The automatic management of drivers and driving spaces', *Geoforum*, 38: 264–275.
Doel, M. A. (1999) *Poststructuralist geographies: the diabolical art of spatial science*, Edinburgh: Edinburgh University Press.
Doherty, B. (1998) 'Opposition to road-building', *Parliamentary Affairs*, 7: 62–75.
——— (1999) 'Paving the way: the rise of direct action against road-building and the changing character of British environmentalism', *Political Studies*, 47: 275–291.
Doherty, B., Paterson, M., Plows, A. and Wall, D. (2002) 'The fuel protests of 2000: implications for the environmental movement in Britain', *Environmental Politics*, 11: 165–173.
——— (2003) 'Explaining the fuel protests', *British Journal of Politics and International Relations*, 5: 1–23.

Domosh, M. (1991) 'Toward an feminist historiography of geography', *Transactions of the Institute of British Geographers*, 16: 95–104.

——— (1996) 'A "feminine" building? Relations between gender ideology and aesthetic ideology in turn-of-the-century America', *Ecumene*, 3: 305–324.

——— (2001) 'The "women of New York": a fashionable moral geography', *Environment and Planning D*, 19: 573–592.

Downs, R. M. and Stea, D. (1974) *Image and environment: cognitive mapping and spatial behavior*, London: Edward Arnold.

——— (1977) *Maps in minds: reflections on cognitive mapping*. [S.l.], Harper and Row.

Dresner, M., Lin, J. S. C. and Windle, R. (1996) 'The impact of low-cost carriers on airport and route competition', *Journal of Transport Economics and Policy*, 30: 309–328.

Drèze, J., Samson, M. and Singh, S. (1997) *The dam and the nation: displacement and resettlement in the Narmada Valley*, Delhi; Oxford: Oxford University Press.

Dutta, A. (2007) *Development-induced displacement and human rights*, New Delhi: Deep and Deep Publications.

Dyck, I., Kontos, P., Angus, J. and McKeever, P. (2005) 'The home as a site for long-term care: meanings and management of bodies and spaces', *Health and Place*, 11: 173–185.

Eade, J. and Garbin, D. (2007) 'Reinterpreting the relationship between centre and periphery: pilgrimage and sacred spatialisation among Polish and congolese communities in Britain', *Mobilities*, 2: 413–424.

Edensor, T. (2001) 'Performing tourism, staging tourism: (re)producing tourist space and practice', *Tourist Studies*, 1: 59.

——— (2002) *National identity, popular culture and everyday life*, Oxford: Berg.

——— (2003) 'M6 – junction 19–16: defamiliarizing the mundane roadscape', *Space and culture*, 6: 151–168.

Ekman, P. (2003) *Emotions revealed: understanding faces and feelings*, London: Weidenfeld and Nicolson.

Elbe, S. (2005) 'AIDS, security, biopolitics', *International Relations*, 19: 403–420.

——— (2008) 'Our epidemiological footprint: the circulation of avian flu, SARS, and HIV/AIDS in the world economy', *Review of International Political Economy*, 15: 116–130.

Evans, J., Hein, J. and Jones, P. (2008) 'Mobile methodologies: theory, technology and practice', *Geography Compass* 2: 1266–1285.

Evans-Pritchard, E. E. (1956) *Nuer religion*, Oxford: Clarendon Press.

Farish, M. (2003) 'Disaster and decentralization: American cities and the Cold War', *Cultural Geographies*, 10: 125–148.

Farnell, B. (1994) 'Ethno-graphics and the moving body', *Man*, 29: 929.

——— (1996) 'Metaphors we move by', *Visual Anthropology*, 8: 311–335.

——— (1999) 'Moving bodies, acting selves', *Annual Review of Anthropology*, 28: 341–373.

Fiske, J. (1989) *Reading the popular*, Boston, Mass.; London: Unwin Hyman.

Foggin, P. M., Foggin, J. M. and Shiirev-Adiya, C. (2000) 'Animal and human health among semi-nomadic herders of Central Mongolia: brucellosis and the bubonic plague in Ovorhangay Aimag', *Nomadic Peoples*, 4: 148–168.

Fortier, A.-M. (2000) *Migrant belongings: memory, space and identity*, Oxford: Berg.

Foucault, M. (2007) *Security, territory, population: lectures at the College de France, 1977–78*, Basingstoke: Palgrave Macmillan.

Francis, G., Fidato, A. and Humphreys, I. (2003) 'Airport-airline interaction: the impact of low-cost carriers on two European airports', *Journal of Air Transport Management*, 9: 267–273.

Friedberg, A. (1993) *Window shopping: cinema and the postmodern*, Berkeley: University of California Press.

Frisby, D. (1985) *Fragments of modernity: theories of modernity in the work of Simmel, Kracauer and Benjamin*, Cambridge: Polity.

Fritzsche, P. (1992) *A nation of fliers: German aviation and the popular imagination*, Cambridge, Mass.: Harvard University Press.

Frykman, J. (1996) *Force of habit: exploring everyday culture*, Lund: Lund University Press; Bromley: Chartwell-Bratt.

Fuller, G. and Harley, R. (2004) *Aviopolis: a book about airports*, London: Blackdog.

Fussell, P. (1980) *Abroad: British literary traveling between the wars*, New York; Oxford: Oxford University Press.

Gagen, E. A. (2006) 'Measuring the soul: psychological technologies and the production of physical health in Progressive Era America', *Environment and Planning D*, 24: 827–850.

Game, A. (2000) 'Falling', *Journal for Cultural and Religious Theory*, 1: 1–41.

——— (2001) 'Riding: embodying the centaur', *Body and Society*, 7: 1–12.

Gatens, M. (1996) 'Through a Spinozist lens: ethology, difference, power', in Patton, P. (ed.) *Deleuze: a critical reader*, Oxford: Blackwells.

Gatrell, A. C. (2005) 'Complexity theory and geographies of health: a critical assessment', *Social Science and Medicine*, 60: 2661–2671.

Gelder, K. and Jacobs, J. M. (1998) *Uncanny Australia: sacredness and identity in a postcolonial nation*, Carlton South, Vic.: Melbourne University Press.

Geschiere, P. and Meyer, B. (1998) 'Globalization and identity: dialectics of flow and closure – introduction', *Development and Change*, 29: 601–615.

Gibbs, A. (2001) 'Contagious feelings: Pauline Hanson and the epidemiology of affect', *Australian Humanities Review*, December 2001.

Gibson, J. J. (1950) *The perception of the visual world*, Cambridge, Mass.: Riverside Press.

——— (1979) *The ecological approach to visual perception*, Dallas; London: Houghton Mifflin.

Giddens, A. (1985) 'Time, space and regionalisation', in Gregory, D. and Urry, J. (eds.) *Social relations and spatial structures*, Basingstoke: Macmillan.

——— (1990) *The consequences of modernity*, Stanford, Calif.: Stanford University Press.

——— (2000) 'Introduction', in Giddens, A. and Hutton, W. (eds.) *On the edge: living wih global capitalism*, London: Jonathan Cape.

Gilroy, P. (1993) *The black Atlantic: modernity and double consciousness*, Cambridge, Mass.: Harvard University Press.

Goetz, A. (2003) *Up, down, across: elevators, escalators and moving sidewalks*, London: Merrell.

Goetz, A. R. (2006) 'Transport geography: reflecting on a subdiscipline and identifying future research trajectories', *Journal of Transport Geography*, 14: 230–231.

Goffman, E. (1961) *Encounters; two studies in the sociology of interaction*, Indianapolis: Bobbs-Merrill.

——— (1963) *Behavior in public places: notes on the social organization of gatherings*, Free Press of Glencoe: New York; Collier-Macmillan: London.

Gordon, M. (1991) *Good boys and dead girls and other essays*, London: Bloomsbury.

Goss, J. (1999) 'Once-upon-a-time in the commodity world: an unofficial guide to mall of America', *Annals – Association of American Geographers*, 89: 45–75.

Gottdiener, M. (2000) *Life in the air: surviving the new culture of air travel*, Lanham, Md.: Rowman and Littlefield.

Graham, B. J. (1995) *Geography and air transport*, Chichester: John Wiley.

Graham, S. (1998) 'The end of geography or the explosion of place? Conceptualizing space, place and information technology', *Progress in Human Geography*, 22: 165–185.

——— (2002) 'Bulldozers and bombs: the latest Palestinian–Israeli conflict as asymmetric urbicide', *Antipode*, 34: 642–649.

——— (2003a) 'Lessons in urbicide', *New Left Review*: 63–78.

——— (2004a) *The cybercities reader*, London; New York: Routledge.

——— (2004b) 'Vertical geopolitics: Baghdad and after', *Antipode*, 36: 12–23.

——— (2005) 'Software-sorted geographies', *Progress in Human Geography*, 29: 562–580.

Graham, S. and Marvin, S. (1996) *Telecommunications and the city: electronic spaces, urban places*, London; New York: Routledge.

——— (2001) *Splintering urbanism: networked infrastructures, technological mobilities and the urban condition*, London; New York: Routledge.

Great Britain. Department of Transport. (1989) *Roads for prosperity*: HMSO.

Gregory, D. (1985) 'Suspended animation: the stasis of diffusion theory', in Gregory, D. and Urry, J. (eds.) *Social relations and spatial structures*, Basingstoke: Macmillan.

——— (2004) *The colonial present: Afghanistan, Palestine, Iraq*, London: Routledge.

Gudis, C. (2004) *Buyways: billboards, automobiles, and the American landscape*, New York; London: Routledge.

Hägerstrand, T. (1967) *Innovation diffusion as a spatial process*, University of Chicago Press.

——— (1982) 'Diorama, path and project', *Tijdschrift Voor Economische en Sociale Geografie*, 73: 323–339.

——— (1985) 'Time-geography: focus on the corporeality of man, society, and environment', in Aida, S. (ed.) *The science and praxis of complexity*, Tokyo: United Nations University.

Haggerty, K. D. and Ericson, R. V. (2000) 'The surveillant assemblage', *British Journal of Sociology*, 51: 605–622.

Haggett, P. (1965) *Locational analysis in human geography*, Edward Arnold: London.

Halfacree, K. (1996) 'Out of place in the country: travellers and the "rural idyll"', *Antipode*, 28: 42–71.

Hall, C. M. (2005) *Tourism: rethinking the social science of mobility*, Harlow: Pearson Education.

Hall, P., Hesse, M. and Rodrigue, J.-P. (2006) 'Editorial: reexploring the interface between economic and transport geography', *Environment and Planning A*, 38: 1401–1408.

Hanlon, N., Halseth, G., Clasby, R. and Pow, V. (2007) 'The place embeddedness of social care: restructuring work and welfare in Mackenzie, BC', *Health and Place*, 13: 466–481.

Hannam, K. (2008) 'Tourism geographies, tourist studies and the turn towards mobilities', *Geography Compass*, 2: 127–139.

Hannam, K., Sheller, M. and Urry, J. (2006) 'Editorial: mobilities, immobilities and moorings', *Mobilities*, 1: 1–22.

Hannerz, U. (1990) 'Cosmopolitans and locals in world culture', in Featherstone, M. (ed.) *Global culture: nationalism, globalization and world culture*, London: Sage.

Hanson, S. (1998) 'Off the road? Reflections on transportation geography in the information age', *Journal of Transport Geography*, 6: 241–250.

Haraway, D. J. (1991) *Simians, cyborgs and women: the reinvention of nature*, London: Free Association.

Hardt, M. and Negri, A. (2000) *Empire*, Cambridge, Mass.: London: Harvard University Press.

Harkness, R. C. (1973) 'Communication innovations, urban form and travel demand – some hypotheses and a bibliography', *Transportation*, 2: 153–193.

Harrington, C. L. and Bielby, D. D. (2005) 'Flow, home, and media pleasures', *Journal of Popular Culture*, 38: 834–854.

Harrison, P. (2007) '"How shall I say it . . .:" relating the nonrelational', *Environment and Planning A*, 39: 590–608.

——— (2008) 'Corporeal remains: vulnerability, proximity, and living on after the end of the world', *Environment and Planning A*, 40: 423–445.

Harvey, D. (1985) 'The geo-politics of capitalism', in Gregory, D. and Urry, J. (Eds.) *Social relations and spatial structures*, London: Macmillan, pp. 128–163.

——— (1989) *The condition of postmodernity: an enquiry into the origins of cultural change*, Oxford: Basil Blackwell.

——— (1996) *Justice, nature and the geography of difference*, Cambridge, Mass.: Blackwell.

——— (2003) *Paris, capital of modernity*, New York; London: Routledge.

——— (2005) *Paris, capital of modernity*, London: Routledge.

Havemann, P. (2005) 'Denial, modernity and exclusion: indigenous placelessness in Australia', *Macquarie Law Journal*, 5: 57–80.

Hayden, D. (1984) *Redesigning the American dream: the future of housing, work, and family life*, New York: W.W. Norton.

Heidegger, M. (1977) *The question concerning technology, and other essays*, New York; London: Harper and Row.

Hein, J. R., Evans, J. and Jones, P. (2008) 'Mobile methodologies: theory, technology and practice', *Geography Compass*, 2: 1266–1285.

Held, D. (1995) *Democracy and the global order: from the modern state to cosmopolitan governance*, Cambridge: Polity.

Hetherington, K. (2000a) 'Museums and the visually impaired: the spatial politics of access', *Sociological Review*, 48: 444–463.

——— (2000b) *New age travellers: vanloads of uproarious humanity*, London: Cassell.

——— (2003) 'Spatial textures: place, touch and praesentia', *Environment and Planning A*, 35: 1933–1944.

Highmore, B. (2005) *Cityscapes: cultural readings in the material and symbolic city*, Basingstoke: Palgrave Macmillan.

Hillier, B. and Hanson, J. (1984) *The social logic of space*, Cambridge: Cambridge University Press.

Hillier, B., Penn, A., Hanson, J. and Grajewski, T. (1993) 'Natural movement: or, configuration and attraction in urban pedestrian movement', *Environment and Planning B*, 20: 29.

Hinchliffe, S. and Bingham, N. (2008) 'Securing life: the emerging practices of biosecurity', *Environment and Planning A*, 40: 1534–1551.

Hindess, B. (2002) 'Neo-liberal citizenship', *Citizenship Studies*, 6: 127–144.

Hine, J. and Mitchell, F. (2001) 'Better for everyone? Travel experiences and transport exclusion', *Urban Studies*, 38: 319–332.

Hirschauer, S. (2005) 'On doing being a stranger: the practical constitution of civil inattention', *Journal for the Theory of Social Behaviour*, 35: 41–67.

Holley, D., Jain, J. and Lyons, G. (2008) 'Understanding business travel time and its place in the working day', *Time and Society*, 17: 27–46.

Holloway, S. L. (2003) 'Outsiders in rural society? Constructions of rurality and nature–society relations in the racialisation of English gypsy-travellers, 1869–1934', *Environment and Planning D*, 21: 695–716.

——— (2005) 'Articulating Otherness? White rural residents talk about gypsy-travellers', *Transactions – Institute of British Geographers*, 30: 351–367.

Holmes, D. (2004) 'The electronic superhighway: Melbourne's CityLink Project', in Graham, S. (ed.) *The cyber cities reader*, London: Routledge.

Hommels, A. (2005) *Unbuilding cities*, Cambridge, Mass.: MIT Press.

Hounshell, D. A. (1984) *From the American system to mass production 1800–1932: the development of manufacturing technology in the United States*, Baltimore, Md.; London: Johns Hopkins University Press.

Howe, S. (2003) 'Edward Said: the traveller and the exile', *Open Democracy*, 1. 10. 2003.

Hoyle, B. S. and Knowles, R. D. (1998) *Modern transport geography*, Chichester; New York: Wiley.

Hua, C.-I. and Porell, F. (1979) 'A critical review of the development of the gravity model', *International Science Reviews*, 2: 97–126.

Hubbard, P. (2006) City, London: Routledge.

Hutchinson, S. (2000) 'Waiting for the bus', *Social Text*, 63: 107–120.

Hyndman, J. (1997) 'Border crossings', *Antipode*, 29: 149–176.

——— (2000) *Managing displacement: refugees and the politics of humanitarianism*, Minneapolis; London: University of Minnesota Press.

Imrie, R. (2000) 'Disability and discourses of mobility and movement', *Environment and planning A*, 32: 1641–1656.

Ingold, T. (2000) *The perception of the environment: essays on livelihood, dwelling and skill*, London: Routledge.

——— (2004) 'Culture on the ground – the world perceived through the feet', *Journal of Material Culture*, 9: 315–340.

——— (2005) 'The eye of the storm: visual perception and the weather', *Visual Studies*, 20: 97–104.

——— (2007a) 'Earth, sky, wind, and weather', *Journal – Royal Anthropological Institute*, 13: 19–38.

——— (2007b) *Lines: a brief history*, London: Routledge.

Isard, W. (1956) *Location and space-economy: a general theory relating to industrial location, market areas, land use, trade and urban structure*, Cambridge, Mass.: MIT Press.

Iyer, P. (2000) *The global soul: jet lag, shopping malls and the search for home*, London: Bloomsbury.

Jackson, J. B. (1984) *Discovering the vernacular landscape*, New Haven, Conn.: Yale University Press.

Jackson, P., Thomas, N. and Dwyer, C. (2007) 'Consuming transnational fashion in London and Mumbai', *Geoforum*, 38: 908–924.

Jacobs, J. (1962) *The death and life of great American cities*, London: Jonathan Cape.

Jain, J. and Lyons, G. (2008) 'The gift of travel time', *Journal of Transport Geography*, 16: 81–89.

Jancovich, M., Faire, L. and Stubbings, S. (2003) *The place of the audience: cultural geographies of film consumption*, London: British Film Institute.
Janelle, D. (1968) 'Central place development in a time–space framework', *Professional Geographer*, 20: 5–10.
——— (1969) 'Spatial reorganization: a model and concept', *Annals of the Association of American Geographers*, 59: 348–364.
——— (1973) 'Measuring human extensibility in a shrinking world', *The Journal of Geography*, 72: 8–15.
Jazeel, T. (2005) 'The world is sound? Geography, musicology and British-Asian soundscapes', *Area*, 37: 233–241.
Jenks, C. and Neves, T. (2000) 'A walk on the wild side: urban ethnography meets the flâneur', *Cultural Values*, 4: 1–17.
Jensen, B. B. (2004) 'Case study Sukhumvit Line – or learning from Bangkok', in Neilsen, T., Albertsen, N. and Hemmersam, P. (eds.) *Urban Mutations: periodization, scale, mobility*, Aarhus: Forlag.
Jensen, O. B. and Richardson, T. D. (2004) *Making European space: mobility, power and territorial identity*, London: Routledge.
Jones, E., Woolven, R., Durodiè, B. and Wessely, S. (2004) 'Civilian morale during the Second World War: responses to air raids re-examined', *Social History of Medicine*, 17: 463–479.
——— (2006) 'Public panic and morale: Second World War civilian responses re-examined in the light of the current anti-terrorist campaign', *Journal of Risk Research*, 9: 57–73.
Jormakka, K. (2002) *Flying Dutchmen: motion in architecture*, Basel; Boston: Birkhäuser.
Kakihara, M. and Sorensen, C. (2002) 'Mobility', *Proceedings of the Annual Hawaii International Conference on System Sciences*: 131.
Kaplan, C. (1996) *Questions of travel: postmodern discourses of displacement*, Durham, NC; London: Duke University Press.
——— (2006) "Mobility and war: the cosmic view of US "air power"', *Environment and Planning A*, 38: 395–407.
Katz, J. (1999) *How emotions work*, Chicago: University of Chicago Press.
Katz, J. E. and Aakhus, M. (2002) *Perpetual contact: mobile communication, private talk, public performance*, Cambridge: Cambridge University Press.
Kaufmann, V. (2002) *Re-thinking mobility: contemporary sociology*, Aldershot: Ashgate.
Kaufmann, V., Bergman, M. M. and Joye, D. (2004) 'Motility: mobility as capital', *International Journal of Urban and Regional Research*, 28: 745–756.
Keeling, D. J. (2007) 'Transportation geography: new directions on well-worn trails', *Progress in Human Geography*, 31: 217–226.
——— (2008) 'Transportation geography – new regional mobilities', *Progress in Human Geography*, 32: 275–284.
Keen, S. (1999) *Learning to fly: trapeze – reflections on fear, trust, and the joy of letting go*, New York: Broadway Books.
Keil, R. and Ali, H. (2007) 'Governing the sick city: urban governance in the age of emerging infectious disease', *Antipode*, 39: 846–873.
Kellerman, A. (2008) 'International airports: passengers in an environment of "authorities"', *Mobilities*, 3: 161–178.

Kelly, R. L. (1992) 'Mobility/sedentism: concepts, archaeological measures and effects.' *Annual Review of Anthropology*, 21: 43–66.

Kenyon, S. (2001) 'Tackling transport-related social exclusion: considering the provision of virtual access to opportunities, services and social networks', *New Technology in the Human Services*, 14: 10–23.

——— (2003) 'Understanding social exclusion and social inclusion', *Proceedings – Institution of Civil Engineers. Municipal Engineer*, 156: 97–104.

Kenyon, S., Lyons, G. and Rafferty, J. (2002) 'Transport and social exclusion: investigating the possibility of promoting inclusion through virtual mobility', *Journal of Transport Geography*, 10: 207–219.

——— (2003) 'Social exclusion and transport in the UK: a role for virtual accessibility in the alleviation of mobility-related social exclusion?', *Journal of Social Policy*, 32: 317–338.

Kern, S. (2003) *The culture of time and space, 1880–1918: with a new preface*, Cambridge, Mass.; London: Harvard University Press.

Kesselring, S. (2006) 'Pioneering mobilities: new patterns of movement and motility in a mobile world', *Environment and planning A*, 38: 269–280.

Kirby, L. (1997) *Parallel tracks: the railroad and silent cinema*, Durham, NC: Duke University Press.

Knowles, R. D. (2006) 'Transport shaping space: differential collapse in time–space', *Journal of Transport Geography*, 14: 407–425.

Knowles, R. D., Shaw, J. and Docherty, I. (2007) *Transport geographies: mobilities, flows and spaces*, Oxford: Blackwell.

Kong, L. and Yeoh, B. S. A. (1997) 'The construction of national identity through the production of ritual and spectacle – an analysis of National Day parades in Singapore', *Political Geography*, 16: 213–239.

Kopomma (2004) 'Speaking mobile: intensified everyday life, condensed city', in Graham, S. (ed.) *The cybercities reader*, London; New York: Routledge.

Kraftl, P. (2007) 'Utopia, performativity, and the unhomely', *Environment and Planning D-Society and Space*, 25: 120–143.

Kraftl, P. and Horton, J. (2008) 'Spaces of every-night life: for geographies of sleep, sleeping and sleepiness', *Progress in Human Geography*, 32: 509–524.

Kwinter, S. (2001) *Architectures of time: toward a theory of the event in modernist culture*, Cambridge, Mass.: MIT Press.

Laban, R. V. (1960) *The mastery of movement*, London: Macdonald and Evans.

Lakoff, G. and Johnson, M. (1980) *Metaphors we live by*, Chicago; London: University of Chicago Press.

Lassen, C. (2006) 'Aeromobility and work', *Environment and Planning A*, 38: 301–312.

Latham, A. (2003) 'Research, performance, and doing human geography: some reflections on the diary-photograph, diary-interview method', *Environment and Planning A*, 35: 1993–2017.

——— (2008) *The Zen of running*, London: University College London.

Latham, A. and McCormack, D. P. (2004) 'Moving cities: rethinking the materialities of urban geographies', *Progress in Human Geography*, 28: 701–724.

Latour, B. (1993) *We have never been modern*, Cambridge, Mass.: Harvard University Press.

——— (1999) *Pandora's hope: essays on the reality of science studies*, Cambridge, Mass.: Harvard University Press.

Latour, B. and Woolgar, S. (1979) *Laboratory life: the social construction of scientific facts*, Beverly Hills: Sage Publications.

Laurier, E. (2001) 'Why people say where they are during mobile phone calls', *Environment and Planning D*, 19: 485–504.

——— (2004) 'Doing office work on the motorway', *Theory Culture and Society*, 21: 261–277.

Law, J. (1994) *Organizing modernity*, Oxford: Blackwell.

——— (2006) 'Disaster in agriculture: or foot and mouth mobilities', *Environment and Planning A*, 38: 227–240.

Law, R. (1999) 'Beyond "women and transport": towards new geographies of gender and daily mobility', *Progress in Human Geography*, 23: 567–588.

LeBreton, D. (2000) 'Playing symbolically with death in extreme sports', *Body and Society*, 6: 1–12.

——— (2004) 'The anthropology of adolescent risk-taking behaviours', *Body and Society*, 10: 1–16.

Lee, J. and Ingold, T. (2006) 'Fieldwork on foot: percieving, routing, socialising', in Coleman, S. and Collins, P. (eds.) *Locating the field: space, place and context in anthropology*, Oxford: Berg.

Lefebvre, H. (2004) *Rhythmanalysis: space, time and everyday life*, London; New York: Continuum.

Lewis, N. (2000) 'The climbing body, nature and the experience of modernity', *Body and Society*, 6: 58–80.

Leyshon, A., Matless, D. and Revill, G. (1995) 'The place of music', *Transactions – Institute of British Geographers*, 20: 423–433.

Licoppe, C. (2004) '"Connected" presence: the emergence of a new repertoire for managing social relationships in a changing communication technoscape', *Environment and Planning D*, 22: 135–156.

Licoppe, C. and Inada, Y. (2006) 'Emergent uses of a multiplayer location-aware mobile game: the interactional consequences of mediated encounters', *Mobilities*, 1: 39–61.

Lindqvist, S. (2007) *Terra nullius: a journey through no one's land*, New York; London: New Press.

Livingstone, K. (2004) 'The challenge of driving through change: introducing congestion charging in central London', *Planning Theory and Practice*, 5: 490–498.

Lloyd, P. E. and Dicken, P. (1977) *Location in space: a theoretical approach to economic geography*, London: Harper and Row.

Lofgren, O. (1999) 'Border crossings: the nationalization of anxiety', *Enthnologica Scandinavia*, 29: 5–27.

Lorimer, H. (2005) 'Cultural geography: the busyness of being "more-than-representational"', *Progress in Human Geography*, 29: 83–94.

Lorimer, H. and Lund, K. (2004) 'Performing facts: finding a way over Scotland's mountains', *Sociological Review*, 52: 130–144.

Lowe, K. (2007) *Inferno: the devastation of Hamburg, 1943*, London: Viking.

Lucretius (1951) *The nature of the universe*, London: Penguin.

Lui, R. (2004) 'The international government of refugees', in Larner, W. and Walters, W. (eds.) *Global governmentality: governing international spaces*. London: Routledge.

Luke, T. and Ó Tuathail, G. (2000) 'Thinking geopolitical space: the spatiality of war, speed, and vision in the work of Paul Virilio', in Crang, M. and Thrift, N. (eds.) *Thinking Space*, London: Routledge.

Lupton, D. (1999) 'Monsters in metal cocoons: "road rage" and cyborg bodies', *Body and Society*, 5: 57–72.

Lury, C. (1997) 'Objects of travel', in Rojek, C. and Urry, J. (eds.) *Touring cultures: transformations of travel and theory*, London: Routledge.

Lyon, D. (2002) 'Surveillance studies: understanding visibility, mobility and the phenetic fix', *Surveillance and Society*, 1: 1–7.

——— (2003a) *Surveillance after September 11*, Cambridge; Malden, Mass.: Polity.

——— (2003b) *Surveillance as social sorting: privacy, risk, and digital discrimination*, London; New York: Routledge.

——— (2007) *Surveillance studies: an overview*, Cambridge; Malden, Mass.: Polity.

Lyons, G., Jain, J. and Holley, D. (2007) 'The use of travel time by rail passengers in Great Britain', *Transportation Research Part A Policy and Practice*, 41: 107–120.

Lyons, G. and Urry, J. (2005) 'Travel time use in the information age', *Transportation Research Part A Policy and Practice*, 39: 257–276.

Lyotard, J. F. (1984) *The postmodern condition: a report on knowledge*, Manchester: Manchester University Press.

Macauley, D. (2002) 'Walking the urban environment: pedestrian practices and peripatetic politics', in Backhaus, G. and Murungi, J. (eds.) *Transformations of urban and suburban landscapes: perspectives*, Lexington: Lexington Books.

McCann, E. J. (2008) 'Expertise, truth, and urban policy mobilities: global circuits of knowledge in the development of Vancouver, Canada's "four pillar" drug strategy', *Environment and Planning A*, 40: 885–904.

MacCannell, D. (1992) *Empty meeting grounds: the tourist papers*, London; New York, NY: Routledge.

McCarthy, A. (2001) *Ambient television: visual culture and public space*, Durham, NC: Duke University Press.

McCormack, D. P. (2002) 'A paper with an interest in rhythm', *Geoforum*, 33: 469–485.

——— (2003) 'An event of geographical ethics in spaces of affect', *Transactions – Institute of British Geographers*, 28: 488–507.

——— (2004) 'Drawing out the lines of the event', *Cultural Geographies*, 11: 211–220.

Macdonald, I. (2005) '"Urban surfers": representations of the skateboarding body in youth leisure', in Caudwell, J. and Bramham, P. (eds.) *Sport, active leisure and youth cultures*, Leisure Studies Association.

McKay, D. (2006) 'Translocal circulation: place and subjectivity in an extended filipino community', *Asia Pacific Journal of Anthropology*, 7: 265–278.

Mackenzie, A. (2006) 'From cafe to parkbench: wi-fi and technological overflows in the city', in Sheller, M. and Urry, J. (eds.) *Mobile technologies of the city*, London: Routledge.

Mackinder, H. J. (1904 [1996]) 'The geographical pivot', in Agnew, J. A., Livingstone, D. N. and Rogers, A. (eds.) *Human geography: an essential anthology*, Oxford: Blackwell.

McLuhan, M. (1964) *Understanding media: the extensions of man*, London: Routledge and Kegan Paul.

McNay, L. (2005) 'Agency and experience: gender as a lived relation', in Adkins, L. and Skeggs, B. (eds.) *Feminism after Bourdieu*, Oxford: Blackwell.

McNeill, W. H. (1995) *Keeping together in time: dance and drill in human history*, Cambridge, Mass.: Harvard University Press.

Maffesoli, M. (1996) *The time of the tribes: the decline of individualism in mass society*, London: Sage.

Makimoto, T. and Manners, D. (1997) *Digital nomad*, Chichester: Wiley.

Malbon, B. (1999) *Clubbing: dancing, ecstasy and vitality*, London: Routledge.

Malkki, L. (1992) 'National geographic – the rooting of peoples and the territorialization of national identity among scholars and refugees', *Cultural Anthropology*, 7: 24–44.

Marcus, G. E. (1995) 'Ethnography in/of the world system: the emergence of multi-sited ethnography', *Annual Review of Anthropology*, 24: 95.

——— (1998) *Ethnography through thick and thin*, Princeton, NJ; Chichester: Princeton University Press.

Marshall, Y. M. (2006) 'Introduction: adopting a sedentary lifeway', *World Archaeology*, 38: 153–163.

Martin, E. (1998) 'Fluid bodies, managed nature', in Braun, B. and Castree, N. (eds.) *Remaking reality: nature at the millenium*. London: Routledge.

Marx, G. T. (1999) 'Measuring everything that moves: the new surveillance at work', *Research in Sociology of Work*: 165–190.

Massey, D. (1993) 'Power-geometry and progressive sense of place', in Bird, J. (ed.) *Mapping the futures: local cultures, global change*, London; New York: Routledge.

——— (1994) *Space, place, and gender*, Minneapolis: University of Minnesota Press.

——— (2005) *For space*, London: Sage.

Massumi, B. (2002) *Parables for the virtual: movement, affect, sensation*, Durham, NC: Duke University Press.

Meade, M. S. and Earickson, R. (2000) *Medical geography*, New York; London: Guilford Press.

Merleau-Ponty, M. (1962) *Phenomenology of perception*, London: Routledge and Kegan Paul.

Merrifield, A. (2008) *The wisdom of donkeys: finding tranquility in a chaotic world*, New York: Walker and Company.

Merriman, P. (2004) 'Driving places: Marc Augé, non-places, and the geographies of england's MI motorway', *Theory Culture and Society*, 21: 145–168.

——— (2005a) 'Materiality, subjectification, and government: the geographies of Britain's Motorway Code', *Environment and Planning D*, 23: 235–250.

——— (2005b) '"Operation motorway": landscapes of construction on England's M1 motorway', *Journal of Historical Geography*, 31: 113–133.

——— (2006a) '"Mirror, signal, manoeuvre": assembling and governing the motorway driver in late 1950s Britain', *Sociological Review*, 54: 75–92.

——— (2006b) '"A new look at the English landscape": landscape architecture, movement and the aesthetics of motorways in early postwar Britain', *Cultural Geographies*, 13: 78–105.

——— (2007) *Driving spaces*, Oxford: Wiley-Blackwell.

Meyrowitz, J. (1985) *No sense of place: the impact of electronic media on social behavior*, New York: Oxford University Press.

Michael, M. (2000) 'These boots are made for walking. . . . mundane technology, the body and human–environment relations', *Body and Society*, 6: 107–126.

Milbourne, P. (2007) 'Re-populating rural studies: migrations, movements and mobilities', *Journal of Rural Studies*, 23: 381–386.

Miller, D. (2001a) *Car cultures*, Oxford: Berg.

——— (2001b) *The dialectics of shopping*, Chicago; London: University of Chicago Press.

Miller, D., Jackson, P. and Thrift, N. (1998) *Shopping, place and identity*, London: Routledge.

Mitchell, K. (2001) 'Transnationalism, neo-liberalism, and the rise of the shadow state', *Economy and Society*, 30: 165–189.

Mitchell, W. J. (1995) *City of bits: space, place, and the infobahn*, Cambridge, Mass.; London: MIT Press.

——— (2004) *M++ the cyborg self and the networked city*, London: MIT Press.

Mocellin, J. and Foggin, P. (2008) 'Health status and geographic mobility among semi-nomadic pastoralists in Mongolia', *Health and Place*, 14: 228–242.

Mokhtarian, P. L. (1990) 'A typology of relationships between telecommunications and Transportation', *Transportation Research Part A policy and Practice*, 24: 231–242.

——— (1991) 'Telecommunications and travel behavior', *Transportation*, 18: 287–289.

——— (2005) 'Travel as a desired end, not just a means', *Transportation Research Part A Policy and Practice*, 39: 93–96.

Morgan, J. (2000) 'To which space do I belong? Imagining citizenship in one curriculum subject', *The Curriculum Journal*, 11: 55–68.

Morgan, N. and Pritchard, A. (2005) 'Security and social "sorting": traversing the surveillance–tourism dialectic', *Tourist Studies*, 5: 115–132.

Morley, D. (2000) *Home territories: media, mobility and identity*, London; New York: Routledge.

Morrill, R. L. (1970) 'Shape of diffusion in space and time', *Economic Geography*, 46: 259–268.

Morris, M. (1988) 'At Henry Parkes Motel', *Cultural Studies*, 2: 1–47.

Morse, M. (1990) 'An ontology of everyday distraction: the freeway, the mall, and television', in Mellencamp, P. (ed.) *Logics of television: essays in cultural criticism*, Bloomington: Indiana University Press.

Mumford, L. (1964) *The highway and the city*, London: Secker and Warburg.

Nash, C. (2000) 'Performativity in practice: some recent work in cultural geography', *Progress in Human Geography*, 24: 653–664.

Nead, L. (2000) *Victorian Babylon: people, streets, and images in nineteenth-century London*, New Haven, Conn.: Yale University Press.

Nemeth, J. (2006) 'Conflict, exclusion, relocation: skateboarding and public space', *Journal of Urban Design*, 11: 297–318.

Neumayer, E. (2006) 'Unequal access to foreign spaces: how states use visa restrictions to regulate mobility in a globalized world', *Transactions – Institute of British Geographers*, 31: 72–84.

Niffenegger, A. (2004) *The time traveler's wife*, London: Random House.

Normark, D. (2006) 'Tending to mobility: intensities of staying at the petrol station', *Environment and Planning A*, 38: 241–252.

Norton, P. D. (2008) *Fighting traffic: the dawn of the motor age in the American city*, Cambridge, Mass.: MIT.

Nyamnjoh, F. B. (2006) *Insiders and outsiders: citizenship and xenophobia in contemporary Southern Africa*, London: Zed.

O'Connell, S. (1998) *The car and British society: class, gender and motoring 1896–1939*, Manchester: Manchester University Press.

Obrador, P. (2003) 'Being-on-holiday: tourist dwelling, bodies and place', *Tourist Studies*, 3: 47–66.

Ohmae, K. (1990) *The borderless world: power and strategy in the interlinked economy*, London: Collins.

Oliver, M. (1990) *The politics of disablement*, London: Macmillan Education.

Olsson, G. (1965) *Distance and human interaction: a review and bibliography*, Philadelphia, PA.: Regional Science Research Institute.

——— (1991) *Lines of power/limits of language*, Minneapolis: University of Minnesota Press.

Ong, A. (1999) *Flexible citizenship: the cultural logics of transnationality*, Durham, NC: Duke University Press.

——— (2006) *Neoliberalism as exception: mutations in citizenship and sovereignty*, Durham, NC; London: Duke University Press.

Packer, J. (2006) 'Becoming bombs: mobilizing mobility in the War of Terror', *Cultural Studies*, 20: 378–399.

Paglen, T. and Thompson, A. C. (2006) *Torture taxi: on the trail of the CIA's rendition flights*, Hoboken, NJ: Melville House.

Parks, L. (2005) *Cultures in orbit: satellites and the televisual*, Durham, NC: Duke University Press.

Pascoe, D. (2001) *Airspaces*, London: Reaktion.

——— (2003) *Aircraft*, London: Reaktion.

Paterson, M. (2000) 'Car culture and global environmental politics', *Review of International Studies*, 26: 253–270.

Peck, J. (2003) 'Geography and public policy: mapping the penal state', *Progress in Human Geography*, 27: 222–232.

Phelan, P. (1993) *Unmarked: politics of performance*, London: Routledge.

Pinder, D. (2004) 'Meanders', in Harrison, S., Pile, S. and Thrift, N. (eds.) *Patterned ground: entanglements of nature and culture*, London: Reaktion.

Pink, S. (2007) 'Walking with video', *Visual Studies*, 22: 240–252.

——— (2008) 'An urban tour: the sensory sociality of ethnographic place-making', *Ethnography*, 9: 175–196.

Pirie, G. H. (2003) 'Cinema and British Imperial Civil Aviation, 1919–1939', *Historical Journal of Film Radio and Television*, 23: 117–132.

Plaut, P. O. (1997) 'Transportation–communications relationships in industry', *Transportation Research Part A Policy and Practice*, 31: 419–429.

Plows, A. (2006) 'Blackwood roads protest 2004: An emerging (re)cycle of UK ecoaction?' *Environmental Politics*, 15: 462–472.

Porter, T. (1997) *The architect's eye: visualization and depiction of space in architecture*, London: E. and F.N. Spon.

Pratt, M.-L. (1986) 'Fieldwork in common places', in Clifford, J. and Marcus, G. E. (eds.) *Writing culture: the poetics and politics of ethnography*, Berkeley; London: University of California Press.

Prince, H. (1977) 'Time and historical geography', in Carlstein, T., Parkes, D. and Thrift, N. J. (eds.) *Making sense of time*, New York: London.
Pritchard, E. E. (1949) *The Sanusi of Cyrenaica*, Oxford University Press: London.
Probyn, E. (1996) 'Queer Belongings: The Politics of Departure', in Grosz, E. and Probyn, E. (eds.) *Sexy bodies: the strange carnalities of feminism*, London: Routledge.
——— (2004) 'Teaching bodies: affects in the classroom', *Body and Society*, 10: 21–44.
Rabinbach, A. (1990) *The human motor: energy, fatigue, and the rise of modernity*, New York: Basic Books.
Raguraman, K. (1997) 'Airlines as instruments for nation building and national identity: case study of Malaysia and Singapore', *Journal of Transport Geography*, 5: 239–256.
Rajan, S. C. (2006) 'Automobility and the liberal disposition', *Sociological Review*, 54: 113–129.
Ravenstein, E. (1889) 'The laws of migration', *Journal of the Royal Statistical Society*, 52: 241–305.
Reason, J. (1974) *Man in motion: the psychology of travel*, London: Weidenfeld and Nicolson.
Relph, E. (1976) *Place and placelessness*, London: Pion.
Reville, G. and Wrigley, N. (2000) 'Introduction', in Reville, G. and Wrigley, N. (eds.) *Pathologies of Travel*. Amsterdam: Rodopi.
Roberts, S., Secor, A. and Sparke, M. (2003) 'Neoliberal geopolitics', *Antipode*, 35: 886–897.
Robertson, S. (2007) 'Visions of urban mobility: the Westway, London', *Cultural Geographies*, 14: 74–91.
Robins, K. (2000) 'Encountering globalization', in Held, D. and McGrew, A. G. (eds.) *The global transformations reader: an introduction to the globalization debate*, Cambridge: Polity Press.
Robinson, J. and Mohan, G. (2002) *Development and displacement*, Milton Keynes: Open University in association with Oxford University Press.
Robinson, M. D. (1998) 'Running from William James' bear: a review of preattentive mechanisms and their contributions to emotional experience', *Cognition and Emotion*, 12: 667–696.
Rose, N. (1996) *Inventing our selves: psychology, power, and personhood*, Cambridge; New York: Cambridge University Press.
Routledge, P. (1994) 'Backstreets, barricades, and blackouts – urban terrains of resistance in Nepal', *Environment and Planning D-Society and Space*, 12: 559–578.
——— (1997a) 'The imagineering of resistance: Pollok Free State and the practice of postmodern politics', *Transactions of the Institute of British Geographers*, 22: 359–376.
——— (1997b) 'A spatiality of resistance: theory and practice in Nepal's revolution of 1990', in Pile, S. and Keith, M. (eds.) *Geographies of Resistance*, London: Routledge.
Rumford, C. (2006) 'Theorizing borders', *European Journal of Social Theory*, 9: 155–170.
——— (2008) 'Introduction: citizens and borderwork in Europe', *Space and Polity*, 12: 1–12.
Said, E. W. (1978) *Orientalism*, Routledge and Kegan Paul: London.
——— (1993) *Culture and imperialism*, London: Chatto and Windus.
Saldanha, A. (2007) *Psychedelic white: Goa trance and the viscosity of race*, Minneapolis: University of Minnesota Press.
Salomon, I. (1985) 'Telecommunications and travel – substitution or modified mobility', *Journal of Transport Economics and Policy*, 19: 219–235.

——— (1986) 'Telecommunications and travel relationships – a review', *Transportation Research Part A-Policy and Practice*, 20: 223–238.

Salter, M. B. (2003) *Rights of passage: the passport in international relations*, Boulder, Co.: Lynne Rienner.

——— (2004) 'Passports, mobility, and security: how smart can the border be?', *International Studies Perspectives*, 5: 71–91.

——— (2006) 'The global visa regime and the political technologies of the international self: borders, bodies, biopolitics', *Alternatives*, 31: 167–189.

——— (2007) 'Governmentalities of an airport: heterotopia and confession', *International Political Sociology*, 1: 49–66.

——— (ed.) (2008) *Politics of/at the airport*, Minneapolis, University of Minnesota.

Santos, G. (2005) 'Urban congestion charging: a comparison between London and Singapore', *Transport Reviews*, 25: 511–534.

Sauer, C. (1952) *Agricultural origins and dispersals*, New York: The American Geographical Society.

Saville, S. J. (2008) 'Playing with fear: parkour and the mobility of emotion', *Social and Cultural Geography*, 9: 891–914.

Scharff, V. (2003) *Twenty thousand roads: women, movement, and the West*, Berkeley; London: University of California Press.

Schivelbusch, W. (1986) *The railway journey: the industrialization of time and space in the 19th century*, Berkeley: University of California Press.

——— (2004) *The culture of defeat: on national trauma, mourning, and recovery*, London: Granta.

Scott, J. C. (1998) *Seeing like a state: how certain schemes to improve the human condition have failed*, New Haven, NJ: Yale University Press.

Seamon, D. (1979) *A geography of the lifeworld: movement, rest and encounter*, London: Croom Helm.

——— (1980) 'Body–subject, time–space routines, and place-ballets', in Buttimer, A. and Seamon, D. (eds.) *The human experience of space and place*, New York: St. Martin's Press.

Sennett, R. (1970) *The uses of disorder: personal identity and city life*, New York: Knopf.

——— (1990) *The conscience of the eye: the design and social life of cities*, New York: Knopf: Distributed by Random House.

——— (1998) *The corrosion of character: the personal consequences of work in the new capitalism*, New York; London: Norton.

Serres, M. (1982) *The parasite*, Baltimore Md.: Johns Hopkins University Press.

——— (1995a) *Angels, a modern myth*, Paris: Flammarion.

——— (1995b) *Genesis*, Ann Arbor: University of Michigan Press.

Serres, M. and Latour, B. (1995) *Conversations on science, culture, and time*, Ann Arbor: University of Michigan Press.

Shanks, M. and Tilley, C. (1993) *Re-constructing archaeology: theory and practice*, London: Routledge.

Sheller, M. (2004a) 'Automotive emotions: feeling the car', *Theory Culture and Society*, 21: 221–242.

——— (2004b) 'Mobile publics: beyond the network perspective', *Environment and Planning D*, 22: 39–52.

——— (2008) 'Gendered Mobilities: epilogue', in Uteng, T. P. and Cresswell, T. (eds.) *Gendered mobilities*, Aldershot: Ashgate.

Sheller, M. and Urry, J. (2000) 'The city and the car', *International Journal of Urban and Regional Research*, 24: 737–757.

——— (2003) 'Mobile transformations of "public" and "private" life', *Theory culture and society*, 20: 107–126.

——— (2004) *Tourism mobilities: places to play, places in play*, London: Routledge.

Shields, R. (1990) 'The logic of the mall', in Riggins, S. H. (ed.) *The socialness of things: essays on the socio-semiotics of objects*, University of Toronto Press: Canada, Berlin.

——— (1991) *Places on the margin: alternative geographies of modernity*, London: Routledge.

Shields, R. and Tiessen, M. (2006) 'New Orleans and other urban calamities', *Space and Culture*, 9: 107–109.

Simmel, G. and Wolff, K. H. (1950) *The sociology of Georg Simmel*, Glencoe, Ill.: Free Press.

Skeggs, B. (2004) *Class, self, culture*, London: Routledge.

Smith, M. P. (2001) *Transnational urbanism: locating globalization*, Malden, Mass.: Blackwell.

——— (2005) 'Transnational urbanism revisited', *Journal of Ethnic and Migration Studies*, 31: 235–244.

Soden, G. (2003) *Falling: how our greatest fear became our greatest thrill: a history*, New York: W.W. Norton.

Solnit, R. (2000) *Wanderlust: a history of walking*, New York: Viking.

——— (2003) *River of shadows: Eadweard Muybridge and the technological wild west*, New York: Viking.

Sparke, M. (2004) 'Passports into credit cards', in Migdal, J. (ed.) *Boundaries and Belonging*, Cambridge: Cambridge University Press.

Spinney, J. (2006) 'A place of sense: a kinaesthetic ethnography of cyclists on Mont Ventoux', *Environment and Planning D*, 24: 709–732.

Steel, C. (2008) *Hungry City*, London: Chatto and Windus.

Stewart, J. Q. (1950) 'The development of social physics', *American Journal of Physics*, 18: 239–253.

Stewart, J. Q. and Warntz, W. (1959) 'Some parameters of the geographical distribution of population', *Geographical Review*, 49: 270–273.

Stewart, K. (2007) *Ordinary affects*, Durham, NC; London: Duke University Press.

Stouffer, S. A. (1940) 'Intervening opportunities: a theory relating mobility and distance', *American Sociological Review*, December: 845–867.

Strathern, M. (1991) *Partial connections*, Rowman and Littlefield.

Swyngedouw, E. (1993) 'Communication, mobility and the struggle for power over space', in Giannopoulos, G. and Gillespie, A. (eds.) *Transport and communications in the new Europe*, London: Belhaven.

Tester, K. (1994) *The flâneur*, London: Routledge.

Thacker, A. (2003) *Moving through modernity: space and geography in modernism*, Manchester: Manchester University Press.

Tharakan, S. (2002) *The nowhere people: responses to internally displaced persons*, Bangalore: Books for Change.

Thomas, D., Holden, L. and Claydon, T. (1998) *The motor car and popular culture in the 20th century*, Aldershot: Ashgate.

Thrift, N. (1983) 'On the determination of social action in space and time', *Environment and planning D: Society and Space*, 1: 23–57.

——— (1990) 'Transport and communications 1730–1914', in Butlin, R. A. and Dodgshon, R. A. (eds.) *An historical geography of England and Wales*, 2nd ed. London: Academic Press.

——— (1996) 'Inhuman geographies: landscapes of speed, light and power', in Thrift, N. (ed.) *Spatial formations*, London: Sage.

——— (1997) 'The Still Point: resistance, expressive embodiment and dance', in Pile, S. and Keith, M. (eds.) *Geographies of resistance*, London: Routledge.

——— (1999) 'Steps to an ecology of place'. in Massey, D. B., Allen, J. and Sarre, P. (eds.) *Human geography today*, Cambridge: Polity Press.

——— (2000a) 'Afterwords', *Environment and Planning D – Society and Space*, 18: 213–255.

——— (2000b) 'Still life in nearly present time: the object of nature', *Body and society*, 6: 34–57.

——— (2004a) 'Driving in the city', *Theory Culture and Society*, 21: 41–59.

——— (2004b) 'Remembering the technological unconscious by foregrounding knowledges of position', *Environment and Planning D*, 22: 175–190.

——— (2006) 'Space', *Theory, Culture and Society*, 23: 139–146.

Thrift, N. and French, S. (2002) 'The automatic production of space', *Transactions – Institute of British Geographers*, 27: 309–335.

Thukral, E. G. (1992) *Big dams, displaced people: rivers of sorrow rivers of change*, New Delhi; London: Sage.

Tiessen, M. (2006) 'Speed, desire, and inaction in New Orleans: like a stick in the spokes', *Space and Culture*, 9: 35–37.

Toffler, A. (1970) *Future shock*, London: Random House.

Tolia-Kelly, D. (2008) 'Motion/emotion: picturing translocal landscapes in the nurturing ecologies research project', *Mobilities*, 3: 117–140.

Tolia-Kelly, D. P. (2004) 'Materializing post-colonial geographies: examining the textural landscapes of migration in the South Asian home', *Geoforum*, 35: 675–688.

——— (2006) 'Mobility/stability: British Asian cultures of landscape and Englishness', *Environment and Planning A*, 38: 341–358.

Tomkins, S. S. and Demos, E. V. (1995) *Exploring affect: the selected writings of Silvan S. Tomkins*, Cambridge: Cambridge University Press.

Tomlinson, J. (1999) *Globalization and culture*, Chichester: Polity Press.

Torpey, J. C. (2000) *The invention of the passport: surveillance, citizenship, and the state*, Cambridge England; New York: Cambridge University Press.

Toynbee, A. (1977) *Mankind and Mother Earth*, London: Book Club.

Toynbee, P. (2000) 'Who's afraid of global culture?', in Giddens, A. and Hutton, W. (eds.) *On the edge: living with global capitalism*, London: Jonathan Cape.

Tuan, Y.-F. (1974) 'Space and place: humanistic perspective', *Progress in Human Geography*, 6: 233–246.

——— (1975) 'Images and mental maps', *Annals of the Association of American Geographers*, 65: 205–213.

——— (1977) *Space and place: the perspective of experience*, London: Edward Arnold.

——— (1978) 'Space, time, place: a humanistic perspective', in Carlstein, T., Parkes, D. and Thrift, N. (eds.) *Timing space and spacing time. Vol. 1*, London: Arnold.

Turnbull, D. (2002) 'Performance and narrative, bodies and movement in the construction of places and objects, spaces and knowledges: the case of the Maltese megaliths', *Theory Culture and Society*, 19: 125–144.

Turner, A. and Penn, A. (2002) 'Encoding natural movement as an agent-based system: an investigation into human pedestrian behaviour in the built environment', *Environment and Planning B*, 29: 473–490.

Turton, D. (2002) 'Refugees and "other forced migrants":Towards a unitary study of forced migration', in Robinson, J. and Mohan, G. (eds.) *Development and displacement*, Milton Keynes: Open University in association with Oxford University Press.

Ullman, E. L. (1957) *American commodity flow: a geographical interpretation of rail and water traffic based on principles of spatial interchange*, University of Washington Press: Seattle.

Urlich, D. U. (1970) 'Introduction and diffusion of firearms in New Zealand 1800–1840', *Journal of the Polynesian Society*, 79: 399–410.

Urry, J. (1990) *The tourist gaze: leisure and travel in contemporary societies*, London; Newbury Park: Sage.

——— (2000) *Sociology beyond societies: mobilities for the twenty-first century*, London; New York: Routledge.

——— (2002) 'Mobility and proximity', *Sociology – the Journal of the British Sociological Association*, 36: 255–274.

——— (2003) *Global complexity*, Cambridge, UK: Polity.

——— (2004) 'Connections', *Environment and Planning D – Society and Space*, 22: 27–37.

——— (2007) *Mobilities*, London: Sage.

Uteng, T. P. and Cresswell, T. (2008) *Gendered mobilities*, Aldershot: Ashgate.

Van Den Abbeele, G. (1992) *Travel as metaphor: from Montaigne to Rousseau*, Minneapolis: University of Minnesota Press.

Van Houtum, H. and Van Naerssen, T. (2002) 'Bordering, ordering and othering', *Tijdschrift Voor Economische En Sociale Geografie*, 93: 125–136.

Vannini, P. (2002) 'Waiting dynamics: bergson, virilio, deleuze, and the experience of global times', *Journal of Mundane Behaviour*, 3: http://www.mundanebehavior.org/issues/v3n2/vannini.htm

Verstraete, G. (2001) 'Technological frontiers and the politics of mobilities', *New Formations*, 26–43.

Vesely, D. (2004) *Architecture in the age of divided representation: the question of creativity in the shadow of production*, Cambridge, Mass.: MIT Press.

Vidal de la Blache, P., Martonne, E. D. and Bingham, M. T. (1965) *Principles of human geography*, London: Constable.

Vigar, G. (2002) *The politics of mobility: transport, the environment, and public policy*, London: Spon Press.

Virilio, P. (2005) *Negative horizon: an essay in dromoscopy*, London: Continuum.

Wacquant, L. C. J. D. (2004) *Body and soul: notebooks of an apprentice boxer*, Oxford; New York: Oxford University Press.

Wajcman, J. (1991) *Feminism confronts technology*, Cambridge: Polity.

Walters, W. (2002a) 'Deportation, expulsion, and the international police of aliens', *Citizenship Studies*, 6: 265–292.

——— (2002b) 'Mapping Schengenland: denaturalizing the border', *Environment and Planning D*, 20: 561–580.

——— (2006) 'Border/control', *European Journal of Social Theory*, 9: 187–203.

Wark, M. (1994) *Virtual geography: living with global media events*, Indianapolis: Indiana University Press.

Watts, L. and Urry, J. (2008) 'Moving methods, travelling times', *Environment and Planning D*, 26: 860–874.

Weizman, E. (2003) 'Strategic points, flexible lines, tense surfaces and political volumes: Ariel Sharon and the geometry of occupation', in Graham, S. (ed.) *Cities, war and terrorism*, Oxford: Blackwell.

——— (2007) *Hollow land: Israel's architecture of occupation*, London: Verso.

Weizmann, E. (2002) 'The politics of verticality', *Open Democracy*, 24/04/2002.

Wellman, B. (2001) 'Physical place and cyberplace: the rise of personalized networking', *International Journal of Urban and Regional Research*, 25: 227–252.

Werbner, P. (1990) *The migration process: capital, gifts and offerings among British Pakistanis*, New York; Oxford: Berg.

——— (1999) 'Global pathways: working class cosmopolitans and the creation of transnational ethnic worlds', *Social Anthropology*, 7: 17–36.

Whatmore, S. (2002) *Hybrid geographies: natures, cultures, spaces*, London; Thousand Oaks, Calif.: Sage.

——— (2003) 'Generating materials', in Pryke, M., Rose, G. and Whatmore, S. (eds.) *Using social theory*, London: Sage.

Whitehead, A. N. (1979) *Process and reality, an essay in cosmology*, New York: Free Press; London: Collier Macmillan.

Whitelegg, J. (1997) *Critical mass: transport environment and equity in the twenty-first century*, London: Pluto.

Wiles, J. (2003) 'Daily geographies of caregivers: mobility, routine, scale', *Social Science and Medicine*, 57: 1307–1325.

Williams, R. (1974) *Television: technology and cultural form*, London: Fontana.

Wohl, R. (1994) *A passion for wings: aviation and the Western imagination, 1908–1918*, New Haven, Conn.: Yale University Press.

——— (2005) *The spectacle of flight: aviation and the Western imagination, 1920–1950*, New Haven, Conn.; London: Yale University Press.

Wolff, J. (1993) 'On the road again: metaphors of travel in cultural criticism', *Cultural Studies*, 7: 224–239.

——— (2006) 'Gender and the haunting of cities (or, the retirement of the flâneur)', in D'Souza, A. and McDonough, T. (eds.) *The invisible flaneuse?: gender, public space and visual culture in nineteenth-century Paris*, London: Routledge.

Wollen, P. and Kerr, J. (2002) *Autopia: cars and culture*, London: Reaktion.

Wolmar, C. (2004) 'Fare enough? The capital has led the way in the UK on innovative transport policies, with the controversial congestion charging zone and a successful push to increase bus use', *Public Finance*, 26–28.

Wood, A. (2003) 'A rhetoric of ubiquity: terminal space as omnitopia', *Communication Theory*, 13: 324–344.

Wood, D. and Graham, S. (2006) 'Permeable boundaries in the Software Sorted Society: surveillance and differentiations of mobility', in Sheller, M. and Urry, J. (eds.) *Mobile technologies of the city*, London: Routledge.

Woolley, H. and Johns, R. (2001) 'Skateboarding: the city as a playground', *Journal of Urban Design*, 6: 211–230.

Worster, D. (1992) *Rivers of empire: water, aridity, and the growth of the American West*, Oxford, England; New York: Oxford University.

Wylie, J. (2002) 'An essay on ascending Glastonbury Tor', *Geoforum*, 33: 441–454.

——— (2005) 'A single day's walking: narrating self and landscape on the South West Coast Path', *Transactions – Institute of British Geographers*, 30: 234–247.

——— (2007) *Landscape*, London: Routledge.

Yantzi, N. M., Rosenberg, M. W. and McKeever, P. (2007) 'Getting out of the house: the challenges mothers face when their children have long-term care needs', *Health and Social Care in the Community*, 15: 45–55.

Yearley, S. (1995) 'Dirty connections: transnational pollution', in Allen, J. and Hamnet, C. (eds.) *A shrinking world?* Oxford: Open University Press.

——— (2000) 'Environmental issues and the compression of the globe', in Held, D. and McGrew, A. G. (eds.) *The global transformations reader: an introduction to the globalization debate*, Cambridge: Polity Press.

Young, I. M. (1990) *Throwing like a girl and other essays in feminist philosophy and social theory by Iris Marion Young*, Bloomington, Ind.: Indiana University Press.

Zelinsky, W. (1973) *A cultural geography of the United States*, Englewood Cliffs, NJ: Prentice-Hall.

Zipf, G. K. (1949) *Human behavior and the principle of least effort: an introduction to human ecology*, Cambridge, Mass.: Addison-Wesley.

INDEX

Aakhus, M. 221
Abler, R. 24, 175, 189
Aboriginal communities, displacement of 115
abstracted mobility 35–6, 135
access to mobility 91–104
Adams, J.S. 24, 189
Adams, Paul 44, 157
Adey, P. 13, 21–2, 184, 200, 206, 208
adventure sports 163
aero-mobilities 177–87; *see also* air travel; airline security; airports
affect 162–73
affections 164–6
affective contagion 192–4
affective moments of clubbing 171–2
Agamben, G. 108
Agar, John 213, 221
agricultural innovation 189–90
Ahmed, Sarah 118, 162, 163, 165–6
AIDS 191
air travel 103; behaviours in 207; rigidities of 183
airline security 218
airports: displacement effects 184; infrastructural complexities of 183; investigations of 93; and non-engagement 207; protests against 127; supporting mobilities 22–3
Ali, S.H. 176, 190–1, 192
Allen, J. 199
Alliez, E. 1
America (Baudrillard) 55
Amin, A. 214
Amoore, Louise 218, 219
Anderson, Jon 161–2, 165
angels and parasites 197–8
Angus, J. 114
Annan, Kofi 1
anthropology 43–4
Appadurai, Arjun 10, 77, 80–1, 187–9, 193–4
approach 12–17
Arcades project (Benjamin) 64
archaeology 43–4
architecture: bodily negotiation of 201; cinematic experiences in 68–9; *see also* built architecture
artefacts, and emotional anchoring 73, 80
artistic movements 6
ascending Glastonbury Tor 158–9
Asian Bird Flu 191
associations, recovery of 161–2
Atkinson, David 60, 118
ATM machines 215
atomized individuals 110–14, 164–5
attachment 71–3, 184–7

attitudinal dispositions to movement 140–1
augmentation 209–21
Auster, Paul 30
auto-mobilities 84–5, 86–8, 103–4, 177–87; *see also* cars
automobilization 89–90

Bachelard, Gason 62–3, 162
Bahnisch, M. 148
Bajc, V. 172
Bale, John 160, 164, 166
Balepur, P. 211
Balzac, Honoré de 65
Bangkok, stratification of travel 96
Bannister, David 164
Barber, Lucy G. 123
Barnes, Trevor J. 48–9, 82
Bartling, H. 86–7
Basel Convention 11
Bassett, K. 162
Bateson, Gregory 149, 171
Batman 30
Baudelaire, Charles P. 62–5, 104
Baudrillard, Jean 55, 169–70
Bauman, Zygmunt 95, 108, 221–2
beaches 154
Bechmann, J. 76
Beck, U. 106, 107
Beckmann, J. 184, 200
behavioural reaction, mobility as 135
behaviouralist approaches to mobility 137–9
Benjamin, Walter 63–5, 66, 68, 83
Bennett, Colin 217, 218
Bergman, Manfred Max 100–1
Bergson, Henri 5–6
Bhabha, H.K. 195–6
Bielby, D.D. 152
big mobilities 9–12
Bigo, D. 218
Bingham, N. 191
bio-centred biomedical monitoring devices 220
biological insecurity 191–2
Bissell, David 205–6, 207–8
Black Atlantic (Gilroy) 195
blockades 129
Blomley, Nick 85, 106, 130–1
Blunt, A. 10, 13, 25, 78
bodies, mediation of 201–3
bodily mobilities, controlling 143–4
body subjects 137–8
Bohm, S. 183
bombs 121–2
Bonsall, P. 103–4
book layout 13–14
boots 204
Borden, Ian 127, 201–2
border control 106–7, 175–6, 191–2, 216
border crossings 108–9
botanizing on the asphalt 64–5
Bourdieu, Pierre 15, 140, 141
Bourke, J. 120, 192
Bowlby, R. 67
boxing 144
Brand, Stuart 27
Braun, B. 191
Brennan, Teresa 166, 168, 192
Brenner, Neil 23, 24
bridges 121–2
briefcases 197
Brown, C. 211
Brown, L.A. 189
Bruno, Giuliana 68, 152, 155–6, 162
BSE 191, 193
Buck-Morss, S. 62, 69
Budd, L. 184
Buildings that Learn (Brand) 27
built architecture: fluidity of 27–8; destruction in war 121–2; and parkour 126–7
Bull, M. 215, 221
bulldozers 121–2
bus stops 113
bus, waiting for 111
Buttimer, A. 138

Calhoun, Craig 106
Canada–US flexible citizenship programme 108–9
Canetti, Elias 169, 170, 171–2

Canzler, W. 5, 13, 35, 102
capacities 164–6
capital: fluidity of 23–4; mobility/mobilities as 74–6, 85, 100–2, 106–7, 188; temporal 99; velocity of circulation 200
capitalism 55–6, 65, 106, 125, 148, 199–200
cars: as buffers 205–6; space of 89; *see also* driver; driving
cartoon comic-book heroes 30
case study boxes 16–17
Casino Royale (Fleming) 117–19, 126
Castells, Manuel 11, 210, 212–13
Castree, N. 13
Cavanaugh, W.T. 172
centrifugal disposition 140–1
centripetal disposition 140–1
Certeau, Michel De 118, 125, 126
Chambers, I. 67
Chang, S.E. 46
Chatty, D. 115
Choo, S. 211
Christaller, Walter 6, 45, 46
Chyba, C.F. 191
cinema 68–9, 117–19, 126, 187, 193–4
cities: rhythm of 30–1; as spatial fixes 23–4; strolling in 63–9
citizenship 105–10; South Africa 55–6
Cityscapes, the matrix (Highmore) 30
Civil Aviation Authority 184
civil defence evacuation schemes 48–9
civil society and participation 88–91
Clarke, Roger 217
Clasby, R. 114
Claydon, T. 178
Cliff, A.D. 190
Cloke, P. 38
closed circuit television (CCTV) 215
Clubbing (Malbon) 171–2
clubbing, affective moments of 171–2
cognitive approaches to mobility 137–9
Cohen, S. 193, 196
Colchester, M. 115
Cold War planning 48–9
Coleman, S. 172
Comaroff, J. 56
comfort, politics of 205–6
commodity chains 9, 53
communal church settings 78, 80–1
communication: and community 166–73; and transport 176–7
community and communication 166–73
commuting 2, 96–9
complementarity 211
Condition of post-modernity (Harvey) 6
connection 198–203
Connell, J. 194–6
Conradson, David 78, 79, 172–3
consumers 63–70
consumerism 106
contagion 189–94
contexts 36–9, 211–14
Cook, I. 53, 71
Cooper, Melinda 191
Corbusier, Charles E.J. Le 49, 81
corridors of travel 180
cosmopolitanism 106–7, 109–10
Coward, Martin 62, 121–2
Crang, M. 51, 53, 93–4, 155
Crary, J. 148, 155
Crawford, M. 68
Cresswell, Tim 4, 21, 22, 34–6, 39, 55–6, 62, 74, 87, 88, 92, 100, 104, 105, 106, 118, 131, 136, 140, 147–8, 193–4, 216
Cromley, E.K. 190
Cronin, A.M. 67
cultural analysis 30–1
cultural diffusion 187–9, 193–6
Culture and Imperialism (Said) 25–6
Cunningham, H. 109
Cwerner, Saulo 96–7
cyborgs 201
cyclical repetition 71–3
cycling 159–62

D'Souza, A. 69
Damasio, A.R. 165
dams and displacements 115–17
dance 134, 142, 147–8, 149, 167, 171

Dant, Tim 201
data flow 218–19
dataveillance 217–18
Davis, Mike 96–7, 113
de Goode, Marieke 218, 219
De Landa, Manuel 18, 168
de-industrialization 75
Death and life of American cities (Jacobs) 139
Debord, Guy 125
degrees of mobility 95
Deleuze, Gilles 8–9, 58–60
DeParle, J. 4
Department for the Environment, Fisheries and Food, UK 9
Department of Homeland Security, US 218
dependencies, public transport provision 113
Der Derian, J. 200
Derudder, B. 180, 183
detachment 63–5
development 209; and displacement 114–17
Development and displacement (Mohan/Robinson) 115
Devriendt, L. 180, 183
Dewsbury, J.D. 27–8, 143, 144
Dhagamwar, V. De 117
diasporas 77–81
Dicken, P. 45, 47–8, 50, 92
difference, politics of 91–104
diffusion 187–96
digital nomadism 62
Dillon, Mick 191
Dimendberg, E. 186, 187
disabled travellers 112, 114
disconnection 203–9
discrimination: migrants 55–6, 109; and mobile surveillance 219
disease, diffusion of 190–2
displacement: and development 114–17; in war 119–20
Distinction (Bourdieu) 140
Dixon, Deborah 193–4
Docherty, I. 179
Dodge, M. 206, 213–14, 215, 218
Doel, Marcus 39
Doherty, Brian 119, 127, 129
doing mobility 134–7
domination, mobility as 117–19
Domosh, M. 25, 67, 69
Dowling, R.M. 13, 78
Downs, R.M. 138
Drakulic, Slaveneka 121
Dresner, M. 183
Dreze, J. 116–17
driver judgements 206
driver-car 201–3
driving 83–4, 163; in India 185–6; and working 180–2
Duncan, Isadora 134, 149
durable familialism 77
Durodie, B. 192
Dutta, A. 115
Dwyer, C. 80
Dyck, I. 114

Eade, J. 172
Earickson, R. 190
Eastern Europe, tourism in 153
ecological perception 150–1
economic distance 46
economic rationality 41, 47–8
Edensor, Tim 155, 185–6, 203
Eisenstein, Sergio 68
Elbe, S. 191
emancipatory significance of travel 67–8
emotion and motion 162–73
emotional anchoring 73, 80
enablers, immobilities as 21–2
entanglements of mobility 105–17
environmental direct action (EDA) 161
Ericson, Richard 218
escapism 62–3, 65
ethnography: mobile 66, 70–1; urban 65–6
ethnoscape 188
Europe, freedom of mobility 109
European business connectivities 182
European Imperialism 190
evacuation, New Orleans 86–7
Evans, J. 161, 162, 219–20
Evans-Pritchard, E.E. 60–1, 133, 135

excavating mobilities 43–4
extraordinary rendition 220
extreme sports 127

Faire, L. 156
falling 37
Falling Down (film) 118
Farish, M. 48–9
Farnell, Brenda 134–5
fast and slow 128–31
fear 165–6
feeling: diffusion of 192–4; of mobility 162–73
feminist theory 24–6
Fidato, A. 183
figures of mobility 39–81
Filipino translocalities 79, 80–1
financescapes 188
Fiske, J. 154
fixity: cities 23-4; and fluidity 69–81; from space to place 73–5
fixtures, mobilities dependence on 213–14
flanêrie: and the arcades 64–5; as mobile ethnography 66; reclarification of 69–70
flâneur 63–9, 168–9
flash mobs 170
Fleming, Ian 117–19
flexibility 182–4
flexible citizenship 106–9
flexible lines 61–2
flight, lines of 57–69
fluidity: of capital 23–4; and fixity 69–81
flying pickets 130–1
Foggin, J.M. 190
follow the people 71
follow the thing 53, 71
food mobilities 9, 53, 191–2; slow food movement 128
foot and mouth disease 190, 193
footwork 196–7
forced displacements 115–20
Fortier, Anne-Marie 78
Foucan, Sebastian 117, 126
Foucault, M. 191
Francis, G. 183
free running *see* parkour
freeways 205
French, S. 214
Friedberg, A. 67, 152, 156
Frisby, D. 27, 65
Fritzche, Peter 185
Frykman, J. 155
Fuller, Gillian 178, 184
Fussell, Paul 68
futurism 6, 63

Gagen, E.A. 168
Game, Anne 37, 205
Garbin, D. 172
Garrison, William 48–9
Gatens, M. 165
Gatrell, A.C. 190
Gelder, K. 115
gendered mobilities 88, 140–1
gendered transport exclusion 110–14
geographical capacity 58
geographical co-presence 28
geographical mobility 37–8
German National Socialist Party 185
Geschiere, Peter 70, 73–4
Gibbs, Anne 192, 193
Gibson, C. 194–6
Gibson, James 150–1
Giddens, Anthony 1, 5, 51, 183, 200
Gilroy, Paul 195
Glastonbury Tor, ascending 158–9
Global complexity (Urry) 21
global cultural economies, landscapes of 188
global positioning satellite systems (GPS) 219
global unity 200
globalization 10–11, 187–9 processes of 95
Globalization (Bauman) 95
go-slow convoys 129
Goetz, A. 4, 178–9
Goffman, Ervin 150, 206–7
Goodwin, M. 38

Gordon, Mary 141
Goss, J. 53
Gottdeiner, Mark 207
Gould, P. 24, 44, 175, 189
Graham, B.J. 183
Graham, Steve 24, 45, 62, 84, 96–7, 122, 209, 211, 212, 213, 219
Grajewski, T. 151, 152
gravity, metaphorical 46–8
Gregory, D. 51, 52, 121
grounds 211–14
Guattari, Felix 8–9, 58–60
Gudis, C. 67
guidebooks 155
gypsies 73

Habermas, Jurgen 89
habits and practices 137–41
habitus 140
Hägerstrand, Torsten 50–2, 97–8, 189
Haggerty, Kevin 218
Haggett, Peter 49, 190
Halfacree, K. 38
Hall, C. Michael 5, 99–100
Hall, P. 179
Halseth, G. 114
Hamnet, C. 199
Hanlon, N. 114
Hannam, K. 5, 92, 102, 153
Hannerz, U. 110
Hanson, J. 151, 152
Hanson, Susan 152, 178–9, 211
Haraway, Donna 25, 26, 201
Harding, Sandra 25
Hardt, M. 59
Harkness, R.C. 211
Harley, Ross 178, 184
Harrington, C.L. 152
Harrison, P. 208
Harvey, David 74–5, 95, 198, 199–200, 213
Haussmann, Georges Eugène 124-5
Havemann, P. 115
Hayden, Dolores 110
Heathrow Airport 185
Heidegger, M. 27
Hein, J.R. 162, 220
Held, D. 107
helicopter travel, stratification of 96–7
Henry Parkes Motel 77
Hesse, M. 179
Hetherington, Kevin 73, 157–8, 172
Hexis 142
Heyman, J.M.C. 109
Highmore, B. 30–1
Hillier, B. 151, 152
Hinchliffe, S. 191
Hindess, B. 105
Hine, Christine 113
Hirschauer, S. 208
Holden, L. 178
Holley, D. 182
Holloway, S.L. 38
Holmes, David 219
home spaces 54; motels as 77
homelessness 88
Hommels, Anique 120–1
Hong Kong, Filipinos in 80–1
horse-as-mediator 204–5
horse-as-transport 198–9
Horton, J. 208
Hounshell, D.A. 38
Howe, S. 26
Hoyle, B.S. 39, 178–9
Hua, C.-I. 47
hub and spoke systems 98, 183
Hubbard, P. 13, 184
humanism 35, 41–2, 138–9
humanistic geographic approach 53–4
Humphreys, I. 183
Hutchinson, Sikivu 111
Hyndman, I. 107

ideology 85–8
ideoscapes 188
idle time 221–3
images: mobility in 134–5; tourist consumption of 153
immigrants, South Africa 55–6
immigration law 107
immobilities, beneficial relationship with mobilities 21

immobility 6–7; and music 196
impaired mobility 86–8
Imrie, Rob 82, 105, 110, 114
Inada, Y. 223
India cabaret (Nair) 187
India, motoring in 185–6
indifference 208
inequality, transport 103–4, 110–14, 184
infinite mobile 50
information communication technologies (ICTs) 209–20
information sharing 218–19
informationalization 217–18
infrastructural moorings 21
Ingold, T. 145–6, 150, 157, 159, 161–2, 164–5, 196–7
innovation diffusion 189–90
Insiders and outsiders (Nyamjoh) 55–6
instinctive mobility 166
interest-excitement affect 192
international security 200
Internet 62, 197, 213, 220
Invention of the passport (Torpey) 105
iPods 215, 221
Isard, Walter 46
Israeli–Palestine conflict: destruction in 122; geometry of 61–2
Iyer, Pico 187

Jackson, John Brinckerhoff 54–5, 72, 89–90
Jackson, P. 69, 80
Jacobs, Jane 115, 139, 143, 169
Jain, J. 182, 221
Jancovich, M. 156
Janelle, D. 199
Jazeel, Tariq 195
Jenks, Chris 66
Jensen, Boris Brorman 96–7
Jensen, O.B. 10, 86, 109
Johns, R. 202
Johnson, M. 14
Jones, C. 183
Jones, E. 192
Jones, P. 162, 220
Jormakka, K. 6
Joye, Dominique 100, 101
Justice, nature and the geography of difference (Harvey) 24, 74–5

Kakihara, M. 210
Kaplan, Caren 59, 120, 200
Kariba Dam, Zimbabwe 116–17
Katja, Franka 5
Katz, J. 202, 209, 221
Kaufmann, Vincent 5, 10, 13, 35, 100–1, 102
Keeling, David 178–9
Keen, Sam 144
Keeping together in time (McNeill) 167–8
Keil, R. 176, 190–1, 192
Kellerman, A. 183
Kelly, C. 103–4
Kelly, R.L. 43–4
Kenyon, S. 88, 89, 91
Kern, S. 6, 198, 200
Kerr, J. 178
Kesselring, S. 5, 13, 35, 101, 102
key ideas boxes 16
Khatmandhu, protests 123–4
kinaesthetic intelligence 144–5
Kingwell, Mark 96
Kitchin, R. 206, 213–14, 215, 218
Klee, Paul 6, 22
Knowles, R.D. 39, 178, 179
Kong, L. 123
Kontos, P. 114
Koolhass, Rem 68–9
Kraftl, P. 126, 208
Kwinter, Sanford 18

Laban, Rudolph 133, 142
labour relations 59–60
Lakoff, G. 14
Land, C. 183
landscapes: driving in 205; effects of warfare 21–2; of global cultural economies 188; interpretation of 157–9; as spatial fixes 23; 204–5
Lassen, Claus 180
Latham, Alan 78, 149, 156, 172–3, 202–3

Latour, B. 71, 146, 191, 197–8
Laurier, Eric 180–1, 220
Law, John 17–18, 190, 192
Law, Robin 113
laws of movement 44–53
Le Breton, D. 163
least effort principle 45
Lee, J. 161–2
Lee, T.Y. 211
Lefebvre, Henri 20–1, 26–7, 29–31
Lewis, N. 127, 158
Leyshon, A. 195
Libya, occupation of 60–1
Licoppe, C. 223
lifts: behaviour in 207; waiting for 208
Lin, J.S.C. 183
Lindqvist, S. 115
'line of site' 152
Lingis, Alphonso 157
little mobilities 6–9
Livingstone, K. 103
Lloyd, P.E. 45, 47–8, 50, 92
Lobo-Guerrero, Luis 191
Lorimer, H. 133–4, 149
Los Angeles, public bus system 111
low-cost airlines 184
Lowe, K. 120
Lucretius 7, 164–5
Lui, R. 108
Luke, T. 58
Lund, K. 149
Lupton, D. 201, 202
Lury, C. 197, 201
Lyon, David 217–18, 219
Lyons, G. 88, 182, 221

McCann, E.J. 5
MacCannell, D. 68
McCarthy, Anne 194
Macauley, D. 125
McCormack, Derek 148–9, 171, 202–3
Macdonald, I. 202
McDonough, T. 69
McKay, D. 79, 80–1
McKeever, P. 114
Mackenzie, A. 38
Mackinder, Halford 57–8
McLafferty, S.I. 190
McLuhan, Marshal 176, 199
McNeill, William 120, 167–8
Madox Ford, Ford 63
Maffesoli, M. 168–72
Makawerekwere 55–6
Makimoto, T. 62
Malbon, Ben 154, 171–2
Malkki, Lisa 39–40
Manners, D. 62
marches 122–5
Marcus, George 70–1
Marshall, Y.M. 43–4
Martin, Emily 166
Marvin, S. 45, 84, 96–7, 209, 212–13
Marx, G.T. 216
Marx, Karl 62, 66
mass displacement 119–20
Massey, Doreen 6, 15, 20, 75–6, 90, 91–3, 95, 103, 104, 108–9
Massumi, Brian 162
material connectivity 190–1
Material culture and national identity (Edensor) 185–6
Matless, D. 195
matter, movements of 8–9
Meade, M.S. 190
meaningful mobilities 34–9
meanings: overview 14: production of 135
mechanical representation 135–6
mediascapes 188
mediated mobile societies 177–87
mediating between 196–209
mediations, overview 16
medical pathology, mobilities as 87–8
memories, recovery of 161–2
mental maps 136
Merleau-Ponty, Maurice 15, 137, 138, 141, 157
Merrifield, A. 205
Merriman, Peter 44, 76, 83, 94, 186–7, 201, 206
metaphors of mobility 39–81
Meyer, Birgit 70, 73–4

Meyrowitz, J. 199
Michael, M. 204
Migrant belongings (Fortier) 78
migrant diasporas 77–81
migrant subjects: emplacing 78; Hong Kong 80–1; South Africa 55–6; *see also* immigration law
migrants, arrivals and departures of 172–3
migration 1–2, 4, 25–6; theory of 46–7
Migration Online 120
Milbourne, P. 38
Miller, D. 69, 178
miner's strike (1980s) 130–1
Mitchell, Julian 113
Mitchell, K. 214
Mitchell, William 199
mobile 'body subject' 138
mobile gaze 193–4
mobile method boxes 16, 17
mobile phones 221–3
mobile practice 150–2
mobile prostheses 220–1
mobile-with 18–19
mobility everywhere 5–12
mobility substitution 211
Mocellin, J. 190
Modern transport geography (Hoyle/Knowles) 39
modernist imaginary 63
modernity, representing/regulating mobility in 147–8
Mohan, G. 115
Mokhtarian, Patricia 211
Molotch, Harvey 28
momentariness 76
Mont Ventoux cycle race 159–60
moorings 20–3
more-than-representational mobilities 137–50, 154
Morgan, J. 58
Morgan, N. 216
Morley, David 33, 38, 95, 96
Morrill, R.L. 190
Morris, Meaghan 77
Morse, Margaret 205
motels 77
motility 100–2
motion and emotion 162–73
mountaineering 149
movement 50–1; laws of 44–53; as mobility without meaning 34–6
moving 150–2
moving on 67, 75–7, 119–22
moving together in time 167–8
multi-sited research 70–1
multimethods 128–9
Mumford, L. 199
music 154, 171–2, 194–6

Nair, Mira 187
Nash, Catherine 147
national identity 80, 185–6
natural vision and movement theory 151
Nature of the universe (Lucretius) 7
nature: engaging with 204; simplifying mobilities to 41–2
Nead, L. 69
Negri, A. 59
Nemeth, J. 202
neoliberalism 55–6, 85–6, 106, 108, 114
network society 11, 212–13
networks 40–56
Neumayer, E. 107
Neves, Tiago 66
new encounters 172–3
new mobilities paradigm 5, 35, 39, 179
New Orleans, evacuation of 86–7
new-age travellers 172
Nexus 108–9
Niffenberger, Audrey 225–7
night-clubs 171–2
nodes 40–56
nomad science 59–61
nomadic attachment 71–3
nomadism: and disease diffusion 190; lines of flight 57–69; negative interpretations of 40–2
non-productive, performance as 142–3
non-representational theory 141–6

non-visual wayfinding 157–8
Normark, D. 76–7
Norton, P.D. 89
Nuer religion (Evans–Pritchard) 134
Nyamjoh, Francis 55–6, 109

O'Connell, S. 178
O'Tuathail, G. 58
Obrador, Pons Pau 154
occupation, resistance to 60–1
Ohmae, Kenichi 67
Oliver, M. 112
Olsson, Gunnar 47, 49–50
On the move (Cresswell) 35, 147–8
Ong, Aiwah 1, 15, 20, 94–5, 106–7, 108
open space 60
openings 150–1
Ord, J.K. 190
Ordinary affects (Stewart) 170
Orientalism (Said) 25–6
Ortis, Elisha 4
Otherness, exposure to 155
out-of-town shopping centres 103
Outline of the theory of practice (Bourdieu) 140
Oyster cards 215–16

Packer, Jeremy 200
Paglen, Trevor 220
pain 160
Panofsky, Erwin 155–6
Paris: *flanêrie* in 64–5; reorganization of 124–5
Park, Robert 42
Parkes, Henry 77
parkour 117–19, 125–7
Parks, L. 58, 193–4
participation and civil society 88–91
Pascoe, D. 62, 119, 184
Passenger Name Record (PNR) agreement (2004) 218
Passport in international relations (Salter) 105
passports 105–6, 218
Paterson, M. 119, 129, 183
path dependency 113
paths 7–8, 50–2
Peck, Jamie 5
Penn, A. 151, 152
performance 137–50; and non-representational theory 141–6
performative orientations 155
petrol crisis protests 129
Phelan, Peggy 142–3
phenomenology 137–8
physically sensed way of being 143
physics of power 117–19
pilgrimage 172
Pink, Sarah 66, 128–9, 161
Pirie, G.H. 39
place: attachment to 71–3; putting on the line 80–1; theory of 54
place taking 19–20
place-making strategies 73–5
placelessness 53–6
places 40–56; progressive 75–7; translocal 77–81
Plows, A. 127, 129
Pnina Plaut 211
Poland, B. 114
politics: of difference 91–104; of comfort 205–6; mobile 117–31; of mobility 85–104; overview 14–15
pollution 11, 184
Porell, F. 47
Porter, T. 69
position, primacy of 93–4
positions 24–6
post-colonial theory 24–6
Pound, Ezra 62
Pow, V. 114
power 58–63, 117–19
power geometries 91–104
practice 137–50
practices: examples of 150–2; overview 15
Pratt, Mary-Louise 146
Prince, Hugh 22
Pritchard, A. 216
Probyn, Elspeth 24–5, 192
production of mobilities 35
productive time 221–3

progressive places 75–7
proprioception 157–9
protest marches 122–5
protests, styles of 123–30
pseudo places 68
psychogeography 125
public mobility infrastructures: growth of 89–90; inequality in 103–4; New Orleans 86–7; planning 48–9; and vertical stratification 96–7
public transport 84; disinvestments in 86–7; financial viability of 103; gendered exclusion 110–14

queer theory 24–6

Rabinbach, A. 147
racial motivation of fear 165–6
Rafferty, J. 88
Raguraman, Kevin 185
railways 200
Rajan, Chella 90
rap 195
Ravenstein, E. 46–7
re-presentation 153
Re-thinking mobility (Kaufman) 10, 100–1
Reason, J. 203–4
rebellion, crushing 124–5
reconstructing mobilities 43–4
Regan, Pricilla 217
regulating mobility 147–8
relative permanencies 75
Relph, E. 54
representation of research 145–6
representing mobility 147–8
research, representation of 145–6
resistance: and marches 122–5; mobility as 117–19; and nomadism 60–1; performance of 126–7
Revill, G. 195
Reville, G. 88
rhythmic patterns 28–31, 148–9, 167–8
Richardson, T.D. 10, 86, 109
Ridge, Tom 106
river systems 49
road congestion charging 103–4
Roads for Prosperity (White Paper) 86
Robbins, Kevin 10
Roberts, S. 107
Robertson, S. 44
Robins, Kevin 195
Robinson, Jenny 115
Robinson, William James 163
Rodrigue, J.-P. 179
Rose, Nikolas 8
Rosenberg, M.W. 114
Rotterdam Kunsthal 68–9
Rousseau, Jean Jacques 160–1, 204
Routledge, Paul 123–4, 127
rules of the road 185–6
Rumford, Chris 5, 10, 91
runner's high 164
running away 163, 165–6
Rwanda, mass displacement 120–1

Said, Edward 25–6
Saldanha, A. 172
Salomon, I. 211
Salter, Mark 105, 107, 108
Samson, M. 116–17
sanctity of place 54–5
Santos, G. 103
São Paulo, helicopter travel in 96–7
SARS (Severe Acute Respiratory Syndrome) 175–6, 190–1
Sauer, Carl 40–1, 189
Saville, S.J. 126–7
scale, as territorial entity 24
Scamon, David 137–9, 169
Scharff, Virginia 25
Schivelbusch, Wolfgang 120, 183, 193, 200, 206
science fiction 30
Scott, James C. 124–5
Secor, A. 107
sedentarism 40–56; in South Africa 55–6
seeing 150–2; tourist gaze 152–62
Sennett, Richard 90, 169
sensing 150–2
Serres, Michel 197–8
sex tourism 187

Shanks, M. 53
Sharon, Ariel 61–2, 122
Shaw, J. 179
Sheller, Mimi 5, 87, 89, 90–1, 92, 102, 153, 162, 163, 180
Shields, R. 69, 87, 154
Shiirev-Adiya, C. 190
shoes 204
Simmel, G. 27, 75, 168
Singh, S. 116–17
site-seeing 152–62
Situationist International (SI) 125
skateboarding and mediation 201–2
Skeggs, Beverly 93–4, 102
skywalk structures 96
slow and fast 128–31
Slow Food and Cittáslow (slow city) movement 128–9
Smith, Michael Peter 78
social as mobility 4
social bonds 166–73
social exclusion 69
social hierarchy 95–7
social identities 184–7
social interaction and mobility 180–2
social justice 91–104
social mobility 37–8, 67, 95, 99–100
social norms 186
social orders, reproduction of 139–41
social physics 45
social spaces, withdrawal from 206–9
sociality, arabesque of 169–70
Sociology beyond societies (Urry) 4
Soden, Garret 47
solidarity 167–8
Solnit, Rebecca 148, 161
Sorensen, C. 210
sound 157–9
South Africa, citizenship 55–6, 109–10
space for time 182
space making 19–20
Space Syntax 151
space, taking hold of 125–7
space–time ballets 139
space–time collage 156
space–time constraints 100
space–time prisms 99
space–time routines 50–2
space to place 73–5
spaces of Indian motoring 185–6
Sparke, Matthew 106, 107, 108, 109
spatial fixes, cities as 23–4
Spatial organization (Abler/Adams/Gould) 44
spatial science and the Cold War 48–9
spectatorship 142–3, 155–6
Spiderman 30
Spinney, Justin 159–60, 220
Splintering urbanism (Graham/Morris) 96–7
stabilities 71–3
Stari Most, Mostar 121–2
Stea, D. 138
Steel, Caroline 9
Stengers, Isabel 145–6
Stewart, James 45, 47
Stewart, Kathleen 170
stimulus–response sequence 138
Stouffer, S.A. 46
strategic points 61–2
Strathern, Marylyn 80
stratification of helicopter travel 96–7
stress, search for 163
strike mobilization 130–1
strolling *see flâneur*
Stubbings, S. 156
surveillance 214–20
Surveillance and society (Bennett/Regan) 217
Swyndegouw, E. 213
symbiosis of mobilities 22–3, 104
symbolic capital 185, 197
synchronicities 26–8
synthesis 69–81
syphilis 88

tactile navigation 158
talking while walking-with 161–2
taste 157–9
technological mobilities 210
technologies, mediation of 196–209
technoscape 188, 193
televisual mobilities 192–4

temporary immobilities 21
territorial conflict 60–2
territorialized space 127
territory, migrant experiences of 79–81; *see also* border control; border crossings
terrorism 200
Tester, K. 65
Thacker, A. 31, 62
Tharakan, S. 116
Thomas, C. 38
Thomas, D. 178
Thomas, N. 80
Thompson, A.C. 220
Thorp, M. 53
Thousand plateaus (Deleuze/Guattari) 59
Thrift, Nigel 1, 69, 100, 120, 134, 136, 140, 143, 146, 164, 176, 177, 200, 201, 206, 209, 214
Throwing like a girl (Young) 141
Thukral, E.G. 116
Tiessen, M. 87
Tilley, Chris 43, 53
Time of the tribes (Maffesoli) 169–70
Time traveller's wife (Niffenberger) 225–7
time: idle and productive 221–3; moving together in 167–8
time–space collage 156
time–space compression 75, 91–3, 198–201
time–space convergence 199–200
time–space distanciation 200
time–space path visualization 52
time–space routines 50–1, 97–9
Toffler, Alvin 89
Tolia-Kelly, D.P. 80
toll roads 103–4
Tomlinson, J. 106
Torpey, John 105
touch 157–9
tourism 1–2
tourist gaze 153
tourists 63–9; use of visual senses 152–62
Toynbee, A. 199
Toynbee, Polly 187
tracing mobilities 219–20
traffic, learning to live with 65
traffic principle 45
trained bodies 144–5
trajectories of movement 26–8
translocal places 77–81
transnationalism 20
transport geographies 178–9
Transport geographies (Knowles *et al.*) 179
transport technology, implications of 103
transport: and communication 176–7; gendered exclusion 110–14; *see also* aero-mobilities; auto-mobilities, public transport
trapeze artists 144
travel data 105–6, 218
travel sickness 203–4
travel time 180–2
Treatise on nomadology (Deleuze/Guattari) 58
Tuan, Yi-Fu 53–4, 71–3, 136
'tunnel effect' 98
Turnbull, D. 43–4
Turner, A. 151
Turton, D. 115–17

Ullman, E.L. 44–5
unrelatedness 208
urban modelling 151–2
Urlich, D.U. 189
Urry, John 4, 5, 8, 9–10, 11, 17, 20–2, 24–5, 28, 39, 62, 68, 89, 90–1, 102, 153, 154–5, 177–8, 180, 182, 183, 184, 192, 200, 201, 221
US Census Bureau 2
Uteng, T.P. 88
utopia 126–7

Van den Abbeele, G. 160–1, 204
van Houtum, H. 108–9
van Naerssen, T. 108–9
Vannini, P. 208
Verma, N. 117
Versey, G.R. 190
Verstraete, Ginette 109

vertical mobility 2, 4, 30, 37–8, 207
vertical stratification 95–7
Vesely, Dalibor 157, 158
Victor Sylvester Dance School 147–8
Vidal de la Blache, Paul 41–2, 46, 189
Vigar, G. 86
Virilio, Paul 120, 198–9, 205–6
virtual mobilities 90–1, 209–14
viruses 175–6, 190–2
visual affordances theory 150–1
visual impairment 157–8
visual mobility 194; and tourism 152–62
vocabulary of mobility 87–8

Wacquant, L.C.J.D. 143, 144–5
waiting 111, 208
Wajcman, Judy 110
walk, the 122–5
walkable surface 151
walking 157–9; rethinking of 149
Wall, D. 129
Walters, William 105, 108
Wanderlust (Solnit) 161
warfare, effects of 119–22
Warntz, W. 47
waste trade routes 11
Watts, L. 17
wayfinding 60, 157–8
Weizman, Eyal 61–2
Wellman, Barry 221, 222
Werbner, P. 79–80
Wessely, S. 192
Whatmore, Sarah 115, 145–6
Whitehead, A.N. 145
Whitelegg, J. 114
Wifi 213
Wiggens, Cynthia 84, 104
Wiles, J. 114
Williams, Raymond 194
Windle, R. 183
Witlox, F. 180, 183
Wittgenstein, Ludwig 136
Wohl, Robert 185
Wolff, Janet 68, 87
Wolff, K.H. 75
Wolff, Virginia 63
Wollen, P. 178
Wolmar, C. 103
women: bodily movements 140–1; difference in mobilities 88; emancipatory significance of travel 67–8; exclusion from *flanêrie* 69; transport exclusion 110–14
Wood, A. 183
Woolgar, S. 71
Woolley, H. 202
Woolven, R. 192
working and driving 180–2
workplace surveillance 216
World Travel and Tourism Council 2
Worster, D. 3
Wrigley, N. 88
Wylie, John 157, 158–9

xenophobia 55–6

Yantzi, N.M. 114
Yearley, S. 11
Yeoh, B.S.A. 123
Young, Iris Marion 141

Zelinsky, Wilbur 37–8
Zipf, George 7–8, 45, 171